T0202182

MASTERS OF THE UNIVERSE

MASTERS OF THE UNIVERSE

Conversations with Cosmologists of the Past

Helge Kragh

Aarhus University

OXFORD
UNIVERSITY PRESS

OXFORD
UNIVERSITY PRESS

Great Clarendon Street, Oxford, OX2 6DP,
United Kingdom

Oxford University Press is a department of the University of Oxford.
It furthers the University's objective of excellence in research, scholarship,
and education by publishing worldwide. Oxford is a registered trade mark of
Oxford University Press in the UK and in certain other countries

Published in the United States of America by Oxford University Press
198 Madison Avenue, New York, NY 10016, United States of America

British Library Cataloguing in Publication Data

Data available

Library of Congress Control Number: 2014936054

ISBN 978–0–19–872289–2

Printed in Great Britain by
Clays Ltd, St Ives plc

To my daughter Line and my son Mikkel, distant relatives of CCN

Preface

Masters of the Universe is about the development of cosmology from around 1910 to the big bang standard model of the mid-1960s. It is unusual in that it presents an important chapter in the history of modern science in a format that is essentially fictitious. The book is based on a series of interview transcripts left by an imagined person (CCN), an engineer and cosmology enthusiast who over half a century conducted interviews with distinguished scientists who in some way or other contributed to the progress of cosmology. While some of the scientists with whom CCN had conversations are recognized as pioneering physicists and astronomers, others are less well known today. The story of CCN and how his material turned up many years after his death is briefly told in the Foreword.

The book includes lightly revised transcripts of thirteen of the interviews that CCN conducted, in one case a joint interview with two physicists. Each interview is introduced by a brief description, taken from CCN's notes, of the circumstances of the interview and the life of CCN at the time it took place. The interview transcript is followed by a brief biography of the interviewee. At the end of each chapter there is a series of comments aimed at clarifying and contextualizing the content. This part includes references to the relevant literature, both primary and secondary sources. The sources are collected in an extensive bibliography at the end of the book.

Despite its unconventional format, mixing authentic history with imagined history, the book is meant as a scholarly contribution to the history of modern cosmology. The interviews are purely fictitious, but they follow the authentic record closely. The form of imagined interviews allows the reader to get an insight into a fuller and, in a sense, more correct version of the history than what many books of the conventional and strictly factual form offer. In addition, by using the interview form the history is presented in a lively and freer way—but also in a person-centred and less systematic way—than in standard history of science. The book is largely non-technical and written for a broad readership with some basic knowledge of physics and astronomy. As indicated by the subtitle, none of the interviewed scientists are alive today (but some of their collaborators are).

Helge Kragh
February 2014

Foreword

My mother had an uncle by the name Carl Christian Nielsen (CCN) about whom little is known. He was born in 1887 and began to study chemical engineering at the Polytechnic College in Copenhagen, but did not graduate. Among his teachers was Einar Biilmann, who is known for his important work in electrochemistry and later became a leading figure in Danish chemistry. I later found out that while he was a chemistry student in Copenhagen CCN also had an absorbing interest in astronomy, which he cultivated as an amateur at a private observatory called the Urania Observatory, located in the Frederiksberg area of Copenhagen. Among those who used the observatory was the astronomer Ejnar Hertzsprung, who was trained as a chemical engineer and was then on his way to a distinguished career in international astronomy. CCN knew Hertzsprung, his senior by 14 years, and it is likely that the latter stimulated his interest in astronomical and cosmological issues.

At the age of 22 CCN went to Germany to complete his education at one of the technical universities, possibly in Karlsruhe. Apparently he did well, for he not only qualified as a chemical engineer but also succeeded in taking out several patents related to industrial catalysis. His father, the owner of a shipping company, died tragically in a car accident in 1912, leaving CCN, his only child, a considerable fortune. CCN stayed in Germany until 1934, after which he migrated to the United States. He never married, although at times he may have come close to doing so (Figure 1). It is not known whether his decision to leave Germany was politically, economically, or personally motivated. Although little is known about his life in the United States, he probably worked as a part-time consultant for chemical companies, mostly on the East Coast. In any case, he was by then a man of independent means, a result mainly of fortunate investments and his lucrative patents. He died in Syracuse, New York, in the summer of 1971. Despite having spent most of his life outside Denmark, he remained a Danish citizen, although in 1942 he also gained United States' citizenship.

Apparently CCN never thought of returning permanently to Denmark and he only went back to his native country on a few occasions. When I was a teenager, aged about 15, there was a kind of family reunion in Copenhagen, where myself and my two brothers met him

Figure 1 My only photograph of CCN, with an unidentified lady and probably from Berlin in the early 1930s. Source: Author's private collection.

for the first and last time. It was a warm summer evening in the Tivoli Gardens, the amusement park in the centre of the city. I remember him only vaguely, an old, reticent, and somewhat eccentric man who spoke a curiously old-fashioned Danish. For some reason I remember that he got quite upset when someone pointed his camera at him. He did not like to be photographed. I also remember that he wore a white stetson, just as we would have expected of a successful Danish-American. But that's about as far as my memory goes.

My mother referred to CCN with some reverence as her "inventor-uncle," but neither she nor others in my family seemed to know much about him, except that he was wealthy and had an interest in all things scientific and technological. Indeed, when my family spoke of him, his money was invariably a subject of conversation. Who would inherit from him? As it turned out, no one would.

In fact, his true interest seems to have been in the development of cosmology, the science of the universe, but I realized that only much later. At the time of his death in 1971 I had recently graduated in physics and chemistry from the University of Copenhagen and forgotten nearly everything about him. After an extended period as a high school teacher I eventually turned to a professional career in history of science. The obscure life of my deceased inventor granduncle was but a vague memory of no relevance. Then, just a few years ago, some distant relatives of mine came into possession of a box with letters and manuscripts that had belonged to him and that for decades had been stored away by one of his American friends. Not knowing what to do with the content of the box, its new owners suggested to me that perhaps it was of some historical value. And indeed it was!

Only after having glanced over the contents did I understand that my granduncle had lived a secret life or, at least had had a secret passion. This passion was to interview distinguished scientists who in some way or other had contributed to the progress of cosmology. Apart from various letters, the main content of the box consisted of a series of carefully written interviews that he had conducted over a period of half a century. The first dated from 1913 and the last from 1965. Four of them are written in hand, the others typed, the last two of them on an electronic typewriter. CCN apparently travelled widely to come into contact with and interview some of the most important astronomers and physicists of his time. And he must have followed the research literature closely for all those years, because the interviews reveal a solid understanding of the problems of mainstream cosmology. No doubt he was a secret cosmology enthusiast, and a most serious one.

Cosmology is one of the relatively few fields of modern science that appeals greatly to amateurs, not only because they find its ambitious scope fascinating but also because some of them believe they have solved "the riddle of the universe." I often receive e-mails and letters from people trying to interest me in this or that speculative cosmological theory that professional scientists prefer to ignore—as I do. However, CCN was not an amateur cosmologist in that dubious sense. There is no indication that he had some pet theory of his own or that he used the interviews to promote unorthodox ideas. He seems to have accepted his own limitations, without any desire to participate in the game of cosmology himself. In other words, his attitude was that of a curious but detached observer, or perhaps a reporter or a historian. His business was not understanding the universe, but understanding the thoughts of those individuals who had unravelled some of the secrets of the mighty universe.

I have sometimes wondered how he, as an unknown engineer, managed to get in contact with these prominent scientists. How did he convince them to take part in an interview? I don't know, but perhaps it was an advantage that he was an outsider, neither a journalist nor a rival scientist, and that the interviews were brief. Moreover, he apparently promised that the interviews would remain confidential and were not aimed for publication. There is also the possibility, not a farfetched one, that he used his wealth to get the interviews through. In other words, that he paid the scientists involved a handsome amount of money. More often than not, scientists are richer in thought than they are in money.

The material I have looked at gives little information about how the interviews were conducted and lacks details about the relationship between the transcripts and the oral conversations. I assume that for some of the interviews conducted after the World War II he used a tape recorder, but I am not sure. Nor can I be sure how faithfully the transcripts mirror the actual interviews. It is evident that when he wrote them down he also edited them, but it is impossible to say when and to what extent. Thus I cannot guarantee the authenticity of the interviews. Given that none of the interviewed scientists are alive today, there is no direct way to confirm that they took place as described. On the other hand, the content is of such a kind that there is no reason to doubt that the conversations actually took place in accordance with the transcripts. For example, some of the scientists refer to personal matters of a kind that CCN could hardly have known about had he not heard them from the scientists themselves.

In other words, the kind of history of science that is presented in the thirteen chapters that follow is quite different from the few examples in which authors have invented fictitious scientists and used these figures to make the past alive. This is what Russell McCormmach, an American historian of science, did in his excellent *Night Thoughts of a Classical Physicist* of 1982, a book that provides a captivating analysis of how an invented German physics professor ("Victor Jacob") responded to the revolutionary trends in early twentieth-century physics. But whereas the story of Victor Jacob was an experiment in imagined historiography, this book is solidly based on authentic sources from the past. Another example of imagined history of science is Peter Bowler's *Darwin . . . off the Record* from 2010 in which he lets an imaginary and anonymous interviewer from the present step back in time to enter a fictional dialogue with Charles Darwin. In this way Darwin is confronted with modern concepts such as genes and chromosomes, much to his surprise. The interviews with

Darwin are based on biographical facts, but of course they are purely fictional, again in contrast to the interviews presented here.

Apart from a few technical papers dating from his youth, CCN seems to have published nothing, and it is unclear what he intended to do with the interviews. He might have thought of turning the interviews into a book, but I find it more likely that he just kept them for his own sake, as a kind of personal treasure. At least this would be in accordance with the picture I formed of this eccentric and enigmatic relative of mine. It is quite clear that for him the interview project was more than just a hobby. It was also a project of enlightenment and edification.

The extant material shows that CCN spoke with 22 astronomers, physicists, and cosmologists, but several of the transcripts he left are incomplete. One of them is with the British-American astronomer Cecilia Payne-Gaposchkin, of which just a single page is missing. I could have included that interview, but decided not to, since it deals primarily with astrophysics and spectroscopy and is only indirectly relevant to cosmology. By the way, it stands out as the only interview CCN conducted with a woman scientist. In what follows, I include the complete interviews in largely the form that CCN wrote them down, except that I have translated the interviews conducted in German and Danish into English. They are only revised very slightly, mainly with regard to language, misprints, and obvious factual errors. The introductory parts, where CCN briefly describes the contexts of the interviews, are unedited. The entire material is now deposited at the Danish State Archive in Copenhagen.

I have provided the interviews with a series of comments with the aim of clarifying the content and making it more understandable. Among other things, I have used the comments to include references to the relevant literature, both primary and secondary sources. For some of the more recent astronomers and physicists the literature includes interviews, and it is interesting to compare them with the older material left by CCN. Moreover, I have supplemented the comments with brief biographical accounts of the interviewed scientists. Altogether there are fourteen of them: Kristian Birkeland (1913), Svante Arrhenius (1913), Karl Schwarzschild (1916), Hugo von Seeliger (1920), Albert Einstein (1928), Willem de Sitter (1933), Georges Lemaître (1938), Arthur Eddington (1938), Edwin Hubble (1951), George Gamow (1956), Fred Hoyle (1958), Hermann Bondi (1958), Paul Dirac (1963), and Robert Dicke (1965). In preparing the book I have been assisted by Line Kragh, who kindly read the manuscript and made a number of useful suggestions.

In my view, the material reported here is of considerable value to the history of science. CCN's choice of interviewees reflects what he at the time considered to be interesting contributions to cosmology, and his choice does not always agree with what later generations would see as important contributions. This is the case in particular with the first three interviews. From a modern perspective Birkeland and Arrhenius scarcely qualify as cosmologists at all, and Schwarzschild's place is more properly in the histories of astronomy and physics. One advantage of CCN's material is that it also focuses on cosmology in those cases where the scientists are primarily known for other work. Apart from Birkeland, Arrhenius, and Schwarzschild, Dirac is a case in point. He was the subject of extensive interviews in 1962–3 by Thomas Kuhn and Eugene Wigner, but these dealt exclusively with his youth and contributions to quantum theory. Among the other interviews I want to highlight is the one with Einstein, dating from a time when he still believed in the static universe, and also the one with de Sitter from 1933, the year before his death. Altogether, the material published here for the first time offers a unique insight into how modern cosmology came into being.[1]

---

[1] My granduncle CCN could have existed, he could have conducted the interviews as described, and the source material could have been deposited at the Danish State Archive. But, as mentioned in the Preface, CCN is a fictional person invented by the author, who is also responsible for the no less fictional interviews. So please don't bother the Danish State Archive with requests of access to the deposited material!

Contents

1

Kristian Birkeland:
From Aurora to the Universe

Interview conducted in the Victoria Hotel, Stockholm, on 8 March 1913. Languages: Danish and Norwegian.

My first two interviews were driven by sheer curiosity and not planned to be the beginning of a series of conversations related to questions in cosmology. Arrhenius and Birkeland were famous Scandinavian scientists and I was fascinated by their ideas, wanting to know more about them and especially their wider cosmological consequences. I had arranged by letter to meet jointly with the two professors, as I had been informed that Birkeland was on a short business trip to the Swedish capital. He had been divorced from his wife a few years earlier and was contemplating leaving Norway and moving to Egypt. (He did so later in 1913, never to return to Norway.) Birkeland and Arrhenius had both agreed to have a joint conversation, and I had prepared a short list of questions that diplomatically took into account their disagreements on some matters, scientific as well as personal. It was well known that although they had common research interests they were also scientific rivals.

The plan was good, but it did not work. On the day of my arrival in Stockholm I was informed by a note in my hotel that Professor Arrhenius found it "inconvenient" to have a joint conversation and preferred to be interviewed alone at some later date. Under the circumstances I had to concur, and fortunately Professor Birkeland had no objections. In fact, I think he was relieved. I felt a little stupid having come to Stockholm, rather than Oslo, to interview a Norwegian physicist. In any case, my first interview took place at the Victoria Hotel in the old town, and it went well.

CCN Professor Birkeland, you are professor of physics in Christiania[1] and internationally known for your work on the aurora or northern lights, as we call it. And you are known in particular for your

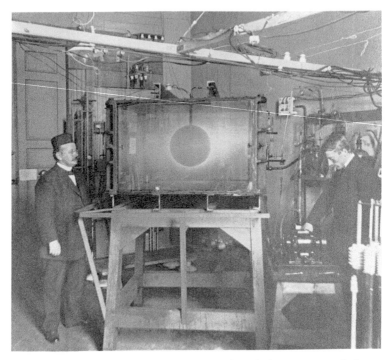

Figure 2 Birkeland (left) and his assistant Karl Devik demonstrating discharge phenomena with the large 36 cm terrella in 1913. Source: Birkeland (1913b, p. 667).

remarkable simulations of auroral and other cosmic phenomena in the laboratory [see Figure 2]. I would like to focus on the wider, more cosmological, aspects of your work, but before doing so could you briefly describe your ideas on the aurora borealis? As far as I understand, these ideas are also at the basis of your thoughts about the structure of the universe.

KB Yes, you are quite right, there is a close connection. Well, we need to go back to 1896, when I was about to return to Christiania from a wonderfully fruitful stay in Geneva and Paris. You see, I examined the effect of magnets on cathode rays, a line of work that had attracted the interest of none other than the great Poincaré.[2] And then it occurred to me that the cathode rays I studied so carefully in my discharge tubes might also be the cause of the aurora that I already knew well. Perhaps, I thought, cathode rays are not confined to the laboratory, but might also exist naturally, in cosmic space.³ Why not? It was a daring thought, but it turned out to be right. Now, the following year Thomson in Cambridge⁴ identified cathode rays as streams of electrons, or what he at the time called corpuscles—by the way, in his famous article on the discovery of the electron he referred to one of my papers, did you know that?

CCN No, I didn't.

KB But he did, oh yes, he did. At any rate, my idea was that the Sun is a huge generator of electrons, and some of the solar electrons are deflected in the Earth's magnetic field near the poles—trapped in it, so to speak. The magnetic storms result from the disturbance of the field, at least in my view. And there's more to it, for the electrons will interact with atoms in the upper rarefied strata of the atmosphere, ionize them, and—*voila!*—cause them to produce the enigmatic auroral light. And then . . .

CCN Sorry to interrupt, but when you say that the aurora is produced by electrons emitted by the Sun reaching the outer atmosphere of the Earth, should it be understood literally? Is it solar electrons that actually hit the molecules in our atmosphere and cause them to emit light?

KB This is what I used to think; but you are right, there are other ways of explaining the action. One might say that the molecules are disturbed by electromagnetic actions that originate in the solar electrons, or that they are hit by ether waves moving with the speed of light rather than by the electrons themselves. Indeed, the latter picture may turn out to be more satisfactory, although from my point of view it doesn't make a great difference. It's just that it is easier to think in terms of electrons.

CCN Thank you. I was curious because I recently read an enjoyable popular book⁵ in which the author suggested that ether waves from the Sun are responsible for the aurora. I have two more questions, just for a start and for clarification. First, with regard to priority, isn't this the same idea that Arrhenius has proposed?⁶ It is often attributed to him, I think. Second, what reasons do we have to believe in the solar cathode ray hypothesis? How can it possibly be verified? After all, we cannot travel to the upper atmosphere and examine what happens up there.

KB I'm glad you are asking me these questions, for Arrhenius only came up with his hypothesis in 1900, several years later than I did, and it is not really the same as mine. In fact, I'm a bit annoyed that he is widely credited for an idea that properly belongs to me. Arrhenius speaks vaguely of electrified dust particles expelled by the Sun and sailing on its radiation pressure, but not of cathode rays carried away from the Sun. His conception of the mechanism causing the aurora, and of the fabric of the universe in general, is certainly ingenious,[7] but it rests entirely on the pressure of starlight. I am not saying that light pressure doesn't exist or doesn't play a role, it certainly does, but it is quite insufficient to explain the cosmic and terrestrial phenomena in question. That should answer your question.

CCN It does, thank you, but there was also this other question of . . .

KB Of course, I forgot. As to your second question it actually relates to the first, for whereas Arrhenius' hypothesis is quite speculative, mine is solidly based on the many experimental analogies that I have demonstrated in my terrella experiments in Christiania.[8] You surely know about them, they are quite famous and in fact unique, for nowhere else have such experiments been made, not in France, Germany, or England, nor for that matter in the United States. Think of it, Norway is a small country that became independent only a few years ago[9] and for centuries was under Danish rule, a kind of colony—and yet we have the only advanced laboratory in the world for the study of solar and other cosmic phenomena. I have, in a sense, brought the celestial bodies down from the heavens and into my laboratory. I think we Norwegians have reason to be proud of ourselves.

CCN Congratulations! To come to the point, just a few weeks ago you gave a lecture to the Norwegian Academy of Science[10] on nothing less than "The Origin of Worlds" that has already attracted international attention. A friend of mine telegraphed me that your lecture was covered in one of the major American newspapers, the *New York Times*. Can you really explain the origin of the universe?

KB No, I cannot, but then I didn't say I could. On the other hand, I think I can explain, if only in principle, how the planets, comets, stars, and nebulae of our universe come into existence—as to the universe as a whole I don't know, it probably makes no sense to speak of its origin. I am currently preparing an article on the subject,[11] but let me try to explain in a few words what I have in mind.

Cosmogony has always been thought to be governed almost entirely by the gravitational force that Newton famously discovered in the seventeenth century. But I have come to the conclusion that electromagnetic processes in space have a role[12] that is equally as important, perhaps even more important. And why do I think this? Well, because my experiments in Christiania with cathode rays and magnetized steel spheres point in this direction. You are right, of course, that I cannot travel to the upper part of the atmosphere or beyond it, but then I don't need to, for I can investigate the heavens in my laboratory.

CCN Fascinating! An electromagnetic universe? But isn't it the force of gravity that governs the behaviour of the celestial bodies?

KB Well, it is, but it's just possible that gravity may ultimately be explained in terms of electromagnetism,[13] as some of my colleagues in physics have attempted to do. But this is not really what I have in mind, for I merely want to supplement Newtonian gravitation with electromagnetic forces, for without these forces we cannot hope to understand the cosmos as a physical machine. You see, the traditional picture of interstellar space is that it is empty, or nearly so, or perhaps that it is filled homogeneously with a kind of subtle ether, but that hardly makes a difference.

At any rate, according to my view this is far from the case. As I see it—and I know it's heretical—we must think of cosmic space as filled with electrons and electrified atoms[14] representing all kinds of chemical elements, atomic particles that make up a tenuous ionized gas that on the average is electrically neutral. The whole of space consists of this shadowy substance charged throughout with electricity, a kind of electrified ether if you prefer, but of the corpuscular kind. I conceive it as a fourth state of matter—in addition to the solid, liquid, and gaseous state—and possibly related to Crookes' so-called radiant matter.[15] It is so important that it ought to have its own name,[16] although I cannot think of one that is both appropriate and apt.

CCN So, this cosmic gas or whatever is invisible to our telescopes, right? Have you formed any idea of how much there is of it, say compared with the ordinary matter that planets and stars are made of?

KB Indeed I have, but only a rough estimate. Let me give an illustration. My calculations indicate that in a spherical volume with a radius equal to the distance to our nearest star, Alpha Centauri, you know, the amount of ionized gas would correspond to one solar mass[17] or one iron atom per eight cubic centimetres of space. You follow me? This corresponds to an average density of the order 10^{-23} g/cm^3, which you may think is such a small number that it scarcely differs from zero, that is, a vacuum. But it is really a large number, and probably even bigger than the average density of stellar matter in the universe. On the other hand, it is small enough not to have an appreciable effect on either the motion of planets or the propagation of starlight in near space. This is of course satisfactory and it gives me reason to believe that it is of the right order of magnitude.

CCN Is that because we know from observations that the planets move as if they travel through empty space, and also that there is no extinction of light in the solar system?

BK Yes, precisely, it's a kind of rarefied dark matter. On the other hand, the density of the gas is large enough to account for the interstellar absorption of light[18] that secures an infinitely large universe. But let me proceed. Let me see . . . , yes, it is also relevant to point out that since the universe as a whole is electrically neutral—at least, we have no reason to think that it's not—the cosmic gas must contain rays of positive electricity in addition to the negative cathode rays. I have made experiments with cathodes made of heavy metals at voltages as high as 20,000 volts and temperatures approaching 2000 degrees centigrade, and they seem to show—but it hasn't been confirmed yet—that the heavy metals can disintegrate into lighter elements. They apparently produce positive particles analogous to alpha rays, only much heavier and moving much faster.

There is probably some kind of extended radioactivity going on in the cosmos, or what I have called "electro-radioactivity," something already anticipated by Rutherford[19] in relation to the origin of the Sun's heat. But it's a new phenomenon that has to be explored further. It's all very exciting, don't you agree?

CCN Absolutely. The kind of electro-radioactivity you speak of must be a subatomic process, a disintegration of the atom itself. How does your idea relate to current theories of atomic structure? Does it presuppose a particular model of the atom?

KB Not really. A model of the kind Thomson proposed[20] will do, but it might also be an atom consisting of electrons and discrete positive charges. What matters is only that the atom can disintegrate into charged particles, and we already know that from ordinary radioactivity. I understand that Rutherford now believes that the positive charge is concentrated in the centre of the atom, what people call the "nucleus." It's probably wrong, but even if it should be the case, my idea would be consistent with it.

CCN I think I follow your line of thought, but . . . if the cosmos is radioactive, such as you and also Rutherford and some other scientists suggest, won't it come to an end—after all, we are told that radioactivity is a strictly irreversible process? It decays over time, but never builds up.

KB I see what you mean, Mr Nielsen, but one must assume that on the average infinite space is in a state of equilibrium. What I mean is that there is equilibrium between disintegration of the heavenly bodies on the one hand, and gathering and condensation of flying electrified corpuscles on the other. The universe is infinite in space, and also in matter and time, the equilibrium process between formation and destruction of matter going on eternally. I'm not the only one to hold a view of this kind. None other than Soddy,[21] who's an authority in the science of radioactivity and a collaborator of Rutherford, has defended somewhat similar ideas based on destructive and constructive atomic processes. He may well agree with my cosmological theory.

CCN And, if I may add, so may Walther Nernst in Germany. He recently argued that radioactive decay in space might result in an ether enriched with energy,[22] which somehow would counteract the universal degradation of matter and energy. I'm not sure if I have understood him, but it occurs to me that his line of thinking is in harmony with what you indicate.

KB Thank you for that information, I'll look up this work of Nernst. Perhaps you have a reference to it?

CCN Not right here, but I shall send it to you as soon as I return to Germany. On the other hand, whether in your version or Nernst's—or Rutherford's for that matter—isn't the idea of radioactive regeneration on a cosmic scale rather speculative? I mean, is there any proof of such processes?

KB I have to admit that there's no proof, at least not so far, but then atomic physics is still in its infancy. At any rate, my laboratory experiments simulating the formation of spiral nebulae and the origin of novae indicate a universe of the kind I have mentioned. I'm a busy man with so many other interests and duties, I unfortunately don't have the time to develop my electromagnetic cosmology into a proper theory, but I hope that others will be inspired to do so.[23]

So, to sum up, I'm convinced that electromagnetism is a key that can unravel many of the secrets of the universe. Not all, but many of them. What is essential to my view, and what is original about it, is that it is a *physical* cosmology guided by experimental facts and analogies. It is not based on gravitation and celestial dynamics alone, like the views of, for example, Seeliger in Germany[24] and the late Poincaré in France, indeed those of all astronomers that I know of.

CCN But then, Professor Birkeland, you are a physicist and not an astronomer. So I wonder, have you discussed your view of the universe with the astronomers? How do they respond to it?

KB They don't, they simply ignore it! But then I must admit that I haven't made any effort to make them interested in my ideas, for example by communicating them in the astronomical journals. It would be a hopeless task trying to convince them, I'm afraid. Astronomers are generally very conservative and quite unable to imagine that there are forces other than gravity operating in the universe. They stubbornly maintain that cosmology is purely an astronomical science—if a science at all—and that physics has no share in it. But really, stellar spectroscopy tells a different story and so, and even more loudly, do my laboratory studies. They ought to be more perceptive to the experimental evidence and not think of astronomy as solely an observational and mathematical science.

CCN Yes, perhaps they should. One final question, if you don't mind. When you are speaking of the universe, what do you mean? It's not an easy concept. To some astronomers the material universe is more or less the same as the Milky Way,[25] while others tend to conceive the spiral nebulae as entire worlds separate from our Milky Way. There's no consensus, as far as I know.

KB There isn't, but at least people agree that the Milky Way is a limited stellar system, and it is hard to imagine that this is all there is. Not only is it hard to imagine, to me it is unimaginable. It is much more satisfactory to think of the universe as indefinitely large and filled throughout with stars, nebulae, and the invisible gaseous "dark matter" that I mentioned.

Now, as to your question of the spiral nebulae and their relation to the Milky Way, there is growing evidence that the Milky Way is itself a spiral nebula and that our solar system is placed near its centre. Many of the nebulae that the astronomers study with their telescopes are stellar systems of the same kind as ours[26]—and their number is indefinitely large. The evidence from my cathode ray experiments cannot prove this picture of the universe, but it adds valuable support to it. There's little doubt that it will be vindicated in the near future.

CCN Professor Birkeland, I hope this is not an inappropriate question, but has it crossed your mind that your work on the aurora and related subjects might be worth a Nobel Prize?

KB I won't deny that it has crossed my mind, and more than just that, but for various reasons a physics prize is unlikely to become a reality. I could mention one particular reason,[27] namely that. . . . No, never mind. On the other hand, I happen to know that I have been nominated for a chemistry prize, so who knows?

CCN Thank you, Professor Birkeland, I will not detain you further. As you kindly reminded me, you are a busy man and on your way to a business meeting at a hotel elsewhere in the city. Just out of curiosity, is it about the inventions you made for the fixation of atmospheric nitrogen[28]?

KB In fact it is, I'm going to meet some potential investors, but enough is enough. You are indeed curious!

Notes

Olaf Kristian B. Birkeland was born in Oslo (then Christiania) on 13 December 1867 and he died in Tokyo on 15 June 1917. Two years after his graduation in physics from the Royal Frederik University in Oslo he went on a scholarship to Paris and Geneva, where his main fields of research were the theory of electromagnetic waves and experiments on cathode rays. His studies of the action of a magnet on the shape of cathode rays led him to discover their so-called magnetic spectrum. It was in this context that he first noticed the analogy between the aurora and the behaviour of cathode rays in a magnetic field, which he further investigated after his appointment in 1898 as professor of physics in Oslo.

He made a series of spectacular laboratory experiments that simulated auroral displays and other phenomena of an astronomical nature (see Figure 2). In addition to his experimental work, he also engaged in extended observations and field research on aurorae. The results were published in a monumental publication, *The Norwegian Aurora Polaris Expedition*, which appeared in two richly illustrated volumes in 1908 and 1913.

Birkeland's work stimulated several young Norwegian scientists, some of them serving as his assistants, to take up research on aurorae and geomagnetism. The most important of them were Carl Størmer, a young mathematician, and Lars Vegard, who in 1918 succeeded Birkeland as professor of physics. Although neither an astronomer nor a cosmologist, in works of 1911–13 Birkeland fearlessly extended his auroral theory to a scenario of the universe and its constituents. His cosmology was innovative by being physical and grounded in electromagnetism (rather than gravitation), but it received only limited scientific recognition. In 1913 he moved from Oslo to Helwan in Egypt, never to return to Norway. He was nominated for a Nobel Prize eight times, but without success.

Apart from his work in pure science, Birkeland was active as an inventor and industrial entrepreneur. He was the holder of 60 Norwegian patents. In 1901 he invented an electromagnetic gun, which was unsuccessful, but a few years later it led him to invent an electric arc device to be used for manufacturing nitrogen fertilizers from atmospheric air. The so-called Birkeland–Eyde method resulted in 1905 in the foundation of Norsk Hydro, which quickly developed into a large and profitable company.

Biographical sources: Jago (2001); Brekke and Egeland (1983, pp. 97–103); Egeland and Burke (2010), with lists of Birkeland's publications and patents.

1. *Christiania*: The capital of Norway was named Christiania, sometimes spelled Kristiania, after the Danish king Christian IV. In 1924 the city officially changed its name to Oslo, a name that was a resurrection of an original one, going back to the Middle Ages.
2. *the great Poincaré*: The French mathematician Henri Poincaré did pioneering work in a variety of subjects, ranging from pure mathematics to electrodynamics, celestial mechanics, and cosmogony. Inspired by Birkeland's experiments on cathode rays, in a paper of 1896 he examined theoretically the motion of an electrical point charge in the field of a (hypothetical) magnetic monopole.
3. *exist naturally, in cosmic space*: One must admit, Birkeland (1896, p. 512) wrote, that "the [cathode] rays come from cosmic space and are in particular absorbed by the Earth's magnetic pole, and that in some way or other they must be attributed the Sun."
4. *Thomson in Cambridge*: J. J. Thomson, the director of the Cavendish Laboratory, is usually credited with having discovered the electron in experiments

on cathode rays in 1897. In his article in the *Philosophical Magazine*, announcing the discovery, he referred to Birkeland's so-called magnetic spectrum of cathode rays (Birkeland 1896). He also adopted his view that the Sun is a generator of electrons, although without acknowledging Birkeland's priority (see Kragh 2013b). The Nobel Prize awarded to Thomson in 1906 was not for his discovery of the electron but more generally for "the great merits of his theoretical and experimental investigations on the conduction of electricity by gases." At about the same time as Thomson, the German physicist and pioneering seismologist Emil Wiechert concluded that cathode rays consist of tiny particles. Wiechert is sometimes mentioned as an independent discoverer of the electron.

5. *I recently read an enjoyable popular book*: Most likely, the enjoyable book was that by Gibson (1911), who on p. 131 writes that one has to "think of the æther waves arriving upon this planet and disturbing sympathetic electrons, causing them to revolve around their atoms in similar fashion to [their] distant fellows who are producing the æther waves." As a result, "they produce that beautiful luminous effect which man describes as an *Aurora*."

6. *the same idea that Arrhenius has proposed*: The auroral and cosmic ideas of Birkeland and Arrhenius are compared and discussed by Kragh (2013b). For more on Arrhenius, see Chapter 2.

7. *is certainly ingenious*: Birkeland admitted that Arrhenius' conception of the universe was "novel and ingenious," but maintained that his own theory was "entirely different, and the probable superiority of [it] is due to the circumstance that more than any other, it is based on experimental analogy." (Birkeland 1913a, p. 32).

8. *my terrella experiments in Christiania*: From 1900 to 1913 Birkeland carried out a series of spectacular experiments in which he placed a steel sphere (a terrella) coated with a phosphorescent substance and provided with an internal electromagnet in a large vacuum chamber. By bombarding the terrella with high-speed cathode rays he created artificial auroral displays, such as luminous rings around the magnetic poles. Some of his other experiments simulated spiral nebulae and the ring system of Saturn. To generate the "solar" cathode rays he made use of an enormous discharge tube, driven by a high-tension generator capable of producing a current of 300 mA at a voltage 15,000 V. For details about Birkeland's experiments, see Rypdal and Brundtland (1997) and Egeland and Burke (2010, pp. 36–43).

9. *a small country that became independent only a few years ago*: Until 1814 Denmark–Norway was a single entity ruled by the Danish king. As a result of the Napoleonic wars, Norway was forced to enter a union with and dominated by Sweden. Norway obtained full independence and its own king when the union was finally dissolved in 1905.

10. *a lecture to the Norwegian Academy of Science*: An extended version of Birkeland's lecture of 31 January appeared in Birkeland (1913a), and it was also published

in Norwegian and French. The *New York Times* of 23 February 1913 covered Birkeland's lecture and his "amazing picture of the future development of the universe" in a long article under the headline "Pictures Universe Electrified Space." (see <http://query.nytimes.com/mem/archive-free/pdf?res=F50 A11FB385F13738DDDAA0A94DA405B838DF1D3>). According to the newspaper, the basis of Birkeland's theory was "the belief that all the suns of the universe were strongly and negatively electrified, their electrical condition being maintained by radiation." Moreover, "the universe is infinite and . . . the whole of space consists of ether charged throughout with electricity."

11. *preparing an article on the subject*: See Birkeland (1913a) and also Birkeland (1913b), a monumental and lavishly illustrated volume reporting observations from a series of auroral expeditions that Birkeland arranged in northern Norway and elsewhere in the Arctic area.

12. *electromagnetic processes in space have a role*: Birkeland's experiments led him to "one cosmogonic theory, in which solar systems and the formation of galactic systems are discussed perhaps rather more from electromagnetic points of view than from the theory of gravitation." (Birkeland 1913b, p. v).

13. *gravity may ultimately be explained in terms of electromagnetism*: There were several attempts in the nineteenth century to explain gravitation as an electrical phenomenon, but they all failed. On the basis of the new electron theory of matter a few physicists, including Hendrik A. Lorentz in the Netherlands and Richard Gans in Germany, derived laws of electrogravitation. However, they were no more satisfactory than earlier theories (see Roseveare 1982, pp. 114–131).

14. *cosmic space as filled with electrons and electrified atoms*: As Birkeland (1913b, p. v) explained: "Space between the heavenly bodies is assumed to be filled with flying atoms and corpuscles of all kinds in such density that the aggregate mass of the heavenly bodies within a limited, very large space would only be a very small fraction of the aggregate mass of the flying atoms there."

15. *Crookes' so-called radiant matter*: The British chemist and physicist William Crookes did pioneering experiments with cathode rays in evacuated tubes, suggesting in the 1890s that the rays consisted of charged particles as a fourth state of matter, an extremely rarefied new kind of ultra-gaseous matter or radiant matter. Although Crookes' view may be seen as an anticipation of Thomson's electron, he did not think of the cathode ray particles as subatomic.

16. *it ought to have its own name*: Without using the name, Birkeland introduced the notion of a space plasma. The term "plasma" for an ionized gas containing about the same number of electrons and positive ions was coined by the American physical chemist Irving Langmuir in 1928. He borrowed the name from the blood plasma carrying with it white and red blood cells. It took a couple of decades until physicists began using the new name.

17. *amount of ionized gas would correspond to one solar mass*: See Birkeland (1913a, p. 20):
    "If, then, we imagine a globe having its center in our solar system and hav-
    ing a radius of five thousand light years, . . . it is possible that the airless space
    in the globe would still contain, in the form of invisible corpuscles, one
    hundred times more material particles than all the stars together."

18. *interstellar absorption of light*: This is a reference to Olbers' paradox, according
    to which the starlight in an infinite universe would make the night sky
    shine bright. The standard solution to the paradox was to assume a tiny
    absorption of light in interstellar space. However, during the first decades
    of the twentieth century there was no agreement among astronomers with
    regard to the question. Harlow Shapley and several other leading astron-
    omers believed space was transparent. The existence of obscuring interstel-
    lar matter was only firmly established in 1930 (see Berendzen et al. 1976,
    pp. 70–99). Birkeland did not use the term "dark matter" in his writings,
    although it would have been an appropriate one.

19. *already anticipated by Rutherford*: Ernest Rutherford was among the physicists
    who, in the first decade of the twentieth century, thought that all atoms
    were radioactive and that the energy of the Sun had its origin in radioactive
    processes unknown on Earth. He speculated that perhaps reverse processes
    of radioactive decay might occur in nature, with heavy atoms being built
    up from lighter and more elementary atoms. Similar speculations were
    entertained by other physicists in the first decade of the twentieth century.
    For the early role of radioactivity in astrophysics and cosmology see Kragh
    (2007a).

20. *A model of the kind Thomson proposed*: In J. J. Thomson's atomic model, electrons
    moved frictionlessly in a positively charged sphere of atomic dimensions.
    The model, or something like it, was still considered a possibility in the
    early 1910s. Rutherford had proposed his nuclear atom in 1911 (without
    using the term "nucleus"), but at the time of the interview few physicists
    outside Manchester found it an attractive alternative.

21. *None other than Soddy*: The British chemist Frederick Soddy, a Nobel laureate
    of 1921, is best known for his discovery of the decay law of radioactivity,
    which he proposed in 1902 in collaboration with Rutherford. In a book of
    1909 he suggested that all matter in the universe "is breaking down and its
    energy being evolved and degraded in one part of a cycle of evolution, and
    in another part still unknown to us, the matter is being built up with the
    utilisation of waste energy. . . . In spite of the incessant changes, an equi-
    librium condition would result, and continue indefinitely." (Soddy 1909,
    p. 241).

22. *an ether enriched with energy*: In a lecture given to the 1912 meeting of the Society
    of German Scientists and Physicians in Münster, the physical chemist
    Walther Nernst imagined "a process which is the reverse of radioactive

decay." On the basis of this hypothesis he suggested that physical processes in the universe might continue indefinitely without violating the second law of thermodynamics (see Kragh 2007b, pp. 214–215).

23. *I hope that others will be inspired to do so*: This did not happen. For a long time Birkeland's ideas of cosmic electromagnetism were ignored. Only much later were they reconsidered within the context of plasma physics and problems of astrophysics. The Swedish physicist Hannes Alfvén, a Nobel laureate of 1970, was instrumental in this process. He considered Birkeland to be the first cosmic plasma physicist and his own work to be in the tradition of Birkeland. For appreciations of Birkeland as the precursor of modern plasma cosmology see Peratt (1985) and Lerner (1991).

24. *Seeliger in Germany*: A reference to the prominent Munich astronomer Hugo von Seeliger, whose view of the universe is the subject of Chapter 4.

25. *more or less the same as the Milky Way*: For all practical purposes, wrote the distinguished American astronomer Simon Newcomb (1906, pp. 5–6), the Milky Way "seems to form the base on which the universe is built and to bind all the stars into a system. . . . As there are no observed facts as to what exists beyond the farthest stars, the mind of the astronomer is a complete blank on the subject."

26. *stellar systems of the same kind as ours*: "I will try to find a new explanation of the formation of the spiral nebulæ and of the origin of the stars known as the Milky Way. There is much support from the view that the Milky Way may be regarded as a spiral nebula, and that spiral nebulæ, as a whole, may be designated as galactic systems." (Birkeland 1913a, p. 21).

27. *I could mention one particular reason*: Birkeland was thinking of Arrhenius, who as a powerful member of the Nobel physics committee used his influence to prevent his rival from Oslo receiving the prize. Birkeland received four nominations for the chemistry prize (1907, 1909, 1912, 1913) and also four for the physics prize (1915–17) (see Egeland and Burke 2010, pp. 138–140).

28. *the fixation of atmospheric nitrogen*: In the Birkeland–Eyde process, dating from 1903, air is led through an electric arc oven at very high temperature, causing nitrogen and oxygen to combine to give nitrogen oxide. By further oxidation and absorption in water, nitric acid is produced and from it calcium nitrate, a fertilizer. Norwegian Birkeland–Eyde factories continued until 1928, by which time they were no longer able to compete with factories based on the alternative Haber–Bosch process in which nitrogen and hydrogen are synthesized under high pressure and by the use of catalysers to ammonia and other nitrogen compounds.

2

Svante Arrhenius' Eternal Cosmos

Interview conducted at the Nobel Institute for Physical Chemistry, Stockholm, on 20 February 1916. Languages: Danish and Swedish.

After my failure to see Arrhenius in 1913 I waited for some time before contacting him again. We finally agreed upon a meeting, but then the European war broke out and for a period of time I could not leave Germany. Many of my friends had volunteered for the army, strongly encouraging me to do the same, but I did not share their patriotic feelings and generally had no desire to be part of the war. Fortunately, as a Danish citizen I was exempted from compulsory service. It took until the beginning of 1916 before I succeeded in meeting Arrhenius. I went to Copenhagen, where I spent most of a week, and then proceeded by boat to a very cold Stockholm. Indeed, I couldn't help thinking of poor Descartes, who suffered terribly from the cold in the Swedish capital and died from pneumonia in 1650, after less than a year in the city. But that is a digression.

The interview took place in Arrhenius' office in the Nobel Institute for Physical Chemistry. It proceeded without problems, as I had hoped. Having conducted the interview I hurried back to Germany, this time directly from Stockholm to Rostock. I needed to prepare for what I had planned should be my next cosmological conversation, with the astronomer Karl Schwarzschild.

CCN Professor Arrhenius [see Figure 3], it is an honour to be able to speak with you and I'm so glad that I have finally managed to see you. Let me see, you received the Nobel Prize in 1903 for your fundamental contributions to physical chemistry, in particular for the ionic theory of dissociation, and you are currently the director of the Nobel Institute for Physical Chemistry placed under the Royal Swedish Academy of Science—and yet your main line of work during the last decade or more has been in other areas of science, am I right?

Svante Arrhenius.

Figure 3 Svante Arrhenius. Source: Zeitschrift für Physikalische Chemie 69 (1909): 1.

SA Well, yes and no, I don't consider myself a chemist any longer, and certainly not in the narrower sense, but rather a physicist working in a broad range of sciences—I'm privileged being able to work on topics for the sole reason that they interest me. But physical chemistry is still an important part of my work. I have done some serious studies relating to the new and fascinating science of immunology, what I call immunochemistry.[1] I coined the name and more or less invented the discipline. And then, of course, I have spent much time on trying to understand the Earth and its surroundings from the perspective of physics—not only its near surroundings, but the entire universe, or as much of it as possible. This is what is sometimes called cosmical physics, and I understand that this is to be the subject of our conversation?

CCN Yes, cosmical physics and even more so cosmology and cosmogony. These are the subjects I'm principally interested in hearing about. You have written a comprehensive textbook on cosmical physics,[2] which is a term that may cause some confusion, so could you briefly explain what it is about?

SA I can, but as you probably know it is not a well-defined subject and people tend to have different conceptions of it. For example, the Austrian school of meteorology[3] thinks of cosmical physics in a sense that is narrower than my conception of it. The basic idea is that, . . . no, let me try again. Cosmical physics is, yes, it's a kind of conglomerate of different fields such as geomagnetism, auroral research, volcanology, oceanography, solar physics, . . . and many more. Climatology is an important part of cosmical physics as well. You may know that I did some work on the effect of carbon dioxide on the temperature of the surface of the Earth.

CCN Yes, now you mention it. That's what you called the hothouse effect,[4] isn't it?

SA Yes, I did, at least I used the expression in some of my later writings. At any rate, I showed that in the future we are likely to have a warmer climate because of the increased amount of carbon dioxide in the atmosphere, due primarily to active volcanoes and the burning of meteorites in the upper parts of the atmosphere. The emission from industry may play part some part as well, but in my view it is a small and temporary one only.

CCN Would the rise in temperature that you foresee be something to worry about?

SA No, not at all. On the contrary, it merely shows that we will not face the problem of a new ice age, which obviously is a good message. But to return to our subject, the broader aim of cosmical physics is to present a *unified* picture of those parts of the earth sciences and the astronomical sciences that can be understood from a *physical* perspective—ultimately as derived from the laws of physics. I admit that this is not a very good definition, but then cosmical physics should not be seen as a scientific discipline in the ordinary sense, but rather as a loose research programme. And as I pointed out in my textbook, to my mind cosmogony is an integral part of cosmical physics—the study of the universe at large belongs to it. But not all agree, in fact I'm pretty much alone in this point of view.

CCN I wonder . . . I have looked at another book on cosmical physics, written by an Austrian by the name Trabert, and he has an interesting section on gravitation and cosmology in which he speculates that space may be curved[5] and the universe therefore is finite. Schwarzschild in Germany had the same idea, even earlier, and he was not the first. I wonder, is that an idea you have contemplated?

SA I once met Trabert, although it was many years ago, and I know his new book. Cosmical physics is a vast subject, but it's cultivated by only a small community of researchers. But to answer your question, I consider ideas of curved space to be a mathematical speculation without the slightest connection to physical reality. My suspicion is that it is an attempt to avoid the conclusion of an infinite universe—a desperate and highly artificial attempt that I cannot take seriously at all. If Trabert believes in it, it cannot be for scientific reasons. He's probably a Catholic.

CCN I see, we'll come back to that. But before we turn to the universe, I noticed that in your textbook you make use of Planck's radiation law, which nowadays is considered fundamental to the new quantum theory. I just wondered, what do you think of this theory which Niels Bohr in Copenhagen has recently used in his amazing model of the structure of atoms—electrons moving in definite orbits around an atomic nucleus? German physicists appear to take it quite seriously, at least some of them. Actually, I tried to contact Bohr when I was in Copenhagen last week, but it turned out that he was in England. I really would have liked to talk with him.

SA I have not followed the literature closely. As far as I am concerned, Planck's law is just a very accurate formula for heat radiation, and therefore a useful one in the area of solar–terrestrial physics, but I tend to find the hypothesis of energy quantization[6] rather fantastic and Bohr's quantum atom no less so. Besides, quantum theory is really of no relevance to what I work with, so I can afford the luxury of ignoring it. It's just not my cup of tea. But I have a student, a Jew,[7] who is interested in these kinds of things, not only in quantum theory but also in statistical mechanics and Einstein's relativity. I advised him to stick to physical chemistry, but he probably won't listen.

CCN No, that's the way students are. Allow me to return briefly to this question of quantum theory, for I recently had the opportunity of listening to a lecture Walther Nernst gave to the German Physical Society[8] in which he suggested empty space to be filled with electromagnetic radiation even at absolute zero temperature. He justified the hypothesis in terms of quantum theory, so it occurs to me that perhaps this theory is not quite as irrelevant with regard to . . .

SA Enough, no reason to go on, I just don't want to comment on Nernst and his apparently wild ideas. You see, we were once friends, but unfortunately[9]. . . . No, never mind, forget about it and let's also forget about quantum theory. We can find more important things to speak about.

CCN No problem, professor. But let me then ask about the role of light pressure in the economy of the universe, such as you described in a most fascinating way in your very successful book[10] nearly a decade ago. I read it as a young man in the original Swedish edition and it made quite an impression on me. As I understand it, the pressure of light, or radiation pressure more generally, plays a key role in your ideas about the structure and life of the cosmos?

SA Absolutely! You see, I actually made use of the light pressure in explaining the aurora and the tails of comets even before Lebedev in Russia demonstrated it experimentally,[11] and then it occurred to me that it really has a much wider use in understanding astronomical phenomena. The basic idea, which I published as early as 1900, is that the Sun pours out into space every second an enormous number of electrified dust particles of different sizes and that the lighter of them are driven far away by its radiation pressure. A tiny fraction of the particles will hit the Earth's atmosphere, but most of them travel much farther away into the abyss of space. And remember, there is nothing special about our Sun, it's just a star, one out of an infinity of them.

CCN Excuse me, but when you say that there is an infinity of stars, do you mean it literally or is it just a way of saying that the number of stars is enormous, that it's beyond imagination?

SA When I say infinity, I mean infinity! Now, the universe is made up of the relatively well-known stars and the much more enigmatic nebulae, and I believe my theory has made the nebulae a little less enigmatic. What do we know about the gaseous nebulae?[12] Well, we know that they are luminous and with an exceedingly low mass density and yet they are bound together by the force of gravity, which means that their brightness cannot be due to the ordinary heat mechanism that causes a gas to glow. No, the outer part of a nebula must be very cold and the light it emits is likely to be due to electrified dust particles expelled by stars close to or far from the nebula. When the particles pass through the outer region they build up an electrical tension, which results in a discharge that renders the gas cloud luminous. It's really much like the light produced by a rarefied gas in a cold discharge tube. A surprising analogy, perhaps, but it's to the point.

CCN Aha, so you have an explanation of the bright but cold nebulae. Does this interaction or energy exchange between nebulae and stars have an even wider significance?

SA It certainly has, and in fact a very important one, for it provides a mechanism ensuring that the clockwork of the universe will never run down.[13] It can't be more important than that, can it? But let me first point out that there are two fundamental questions in cosmology, the one relating to the size of the universe and the other to its duration in time. These are old questions, but they are still with us and I'm convinced that the correct answer to both of the questions can be given in one word: *infinity*.

CCN It's a terrifying concept, at least that's how I feel about it. Could you please elaborate?

SA It's not terrifying at all, not at all. Not only is the universe infinite and uniformly populated with stars and nebulae, it has also existed eternally and, on a large scale, in roughly the same form[14] that we observe it today. There is really nothing new in that, although there are people who consider it controversial, even impious. I am aware of various objections to the infinite stellar universe, quite apart from those who are religiously motivated[15] and whom I cannot take seriously at all. But they fail to impress me. Some twenty years ago Seeliger in Munich came up with an objection[16] that was clever but, to my mind, inconclusive. I can give you my own arguments, if you want, but it's probably easier to look them up in my writings. I think you'll find them convincing.

CCN Thank you, I will look them up when I return. In fact, I'm planning an interview with Professor Seeliger, perhaps next year, and then I will know more about it.

SA I wish you luck with Seeliger, he has a reputation for having a temper. . . . You may think that my belief in the infinite universe is just a faith, which it is perhaps—I mean, it cannot be proved by observation. But I can assure you that this "faith" of mine is scientifically grounded. I hope that is clear.

CCN Quite clear. I take your word for it, professor. Your "faith," to use the expression, seems to be shared by Professor Birkeland, with whom I had a conversation three years ago. It was here in Stockholm, as you may remember. Are you aware of his ideas about the universe?

SA I have been told about them, but see no reason to study his writings.[17]

CCN But wouldn't you agree that . . .

SA No, I wouldn't! Birkeland speculates instead of arguing scientifically. His ideas are based on superficial analogies only, and it is characteristic that he hasn't published his work in any of the recognized astronomical or astrophysical journals. There are reasons for it, of course. Shall we proceed?

CCN As you wish. Now, allow me to quote from one of your papers[18] dealing with the infinite universe, in which you conclude as follows: "A finite universe in infinite space, or a universe in which matter is infinitely diluted, cannot have existed for an unlimited time, and consequently it does not correspond to our experience regarding the properties of energy and matter." What I do not quite understand is the word "consequently," for isn't there the possibility, at least from a logical point of view, that the universe has existed for a finite time only? That somehow it came into existence, was created?

SA Hmm, perhaps from a logical point of view, but certainly not from a scientific point of view. You see, our entire understanding of nature is based on the fundamental principles of the indestructibility of energy and of matter.[19] But if the universe somehow came into existence[20] a finite period of time from now, or if it ceases to exist a finite time from now, these principles are violated and we have stopped doing science. It can't be otherwise. The notion of a temporally finite universe belongs to religion and superstitious belief, it's simply unscientific— one cannot even comprehend what a beginning of or an end to the universe should mean. I tend to find it questionable whether it can be defended even from what you call a logical point of view.

My own theory is not concerned with creation at all, although it is very much concerned with *transformation*, from one form of matter to another and from one form of energy to another. It's a theory of how nebulae originate from stars and stars from nebulae, but these and other cosmic processes occur continually as kinds of huge waves or oscillations in endless space and time. Creation in the literal sense plays no part at all. For this reason the English title of my book is slightly misleading. I tried to persuade the American publisher to choose a title other than *Worlds in the Making*, but they insisted that it was apt and would help to sell the book.

CCN Yes, you make your point very clearly in your book, where you say that violation of energy conservation is incomprehensible. On the other hand, doesn't the history of science tell another story? After all, there was a time not so long ago when scientists could easily comprehend energy non-conservation. Only after the 1850s or so did it become clear that . . .

SA Well, hmm, of course, I know the history of science very well, thank you. What I mean is that to a *modern* scientist the conservation of energy is a necessary truth, and to him a violation of energy conservation on a cosmic scale is doubly incomprehensible, simply outrageous.

CCN You may well be right, but critics might still argue, as indeed they do, that while you keep to the principles of energy and matter conservation you do it at the cost of violating another fundamental principle of science, the second law of thermodynamics. Isn't there a problem?

SA Only on the face of it, I'll say, but not if one goes deeper. Clausius' law[21] is indeed important, but in its standard formulation it cannot be perfectly true or valid on a cosmological scale, for then it would have the absurd consequence that a scientific law leads to a denial of science.[22] You see what I mean? Fortunately there are ways to avoid the absurdity, and that's where the pressure of light comes in. One must expect that something similar to Maxwell's demon[23] operates in the nebulae, only on a hugely greater scale. As I said, the nebulae absorb radiation energy and dust particles from the stars, but the point is that they do it in such a way that they remain cold and act as storage houses for the matter and energy emitted by the stars. To put it briefly, although entropy increases in the hot stars, in agreement with Clausius, it *decreases* in the nebulae, and the overall result may well be that on the average the amount of entropy in the universe remains the same.

I know that this explanation has been met with scepticism, but there *must* be some counter-entropic cosmic mechanism, and the one I have proposed is at least a possible one. Poincaré objected to what he called "Arrhenius' demon,"[24] and so I wrote him a letter in which I suggested a revision of the hypothesis. I don't know if he found my response satisfactory, for he was ill and died shortly later. Should I be proud to have a demon named after me? I'm not sure.

CCN It's new to me that you corresponded with Poincaré on this matter. But he's not the only sceptic. I saw in the recent edition of Newcomb–Engelmann[25] that Schwarzschild isn't happy with your hypothesis either.

SA Yes, I know; he came up with a clever and rather severe criticism, I must admit. Perhaps there was something wrong with the mechanism I suggested, but then there must be other mechanisms that lead to the same result and counter the growth in entropy. The conclusion cannot possibly be wrong.

CCN Well, there is one more aspect of your book on the making of worlds that I hope we can cover and which has attracted much public attention—I'm referring of course to your daring suggestion of organic life in the universe.

SA Okay, let this be the last topic of our conversation. By the way, it's unnecessary to speak of "organic life," as you do. Life *is* organic, by definition, how could it possibly be inorganic? Well, I actually made the suggestion as early as 1901,[26] but it was only with my book it became widely known.

You cannot imagine how many letters I have received on this topic! Many of the letters and newspaper articles describe my theory as being about "the origin of life," but that's a complete misunderstanding, for I don't believe that life has an origin in any absolute sense. I tend to consider life as an elementary quality that has always existed and which may even be governed by some as yet unknown conservation law, not unlike the laws governing matter and energy. No, my theory is about how primitive life forms—what I call "panspermia,"

I can't remember if I coined the name[27] or borrowed it from someone else—well, it's about how these spores or seeds of life are propelled through space by the action of the stars' radiation pressure. So it's an attempt to explain how life came to Earth, after which evolution took over. But life cannot be restricted to the Earth, there must be life all over the universe, if not highly developed life forms. I don't believe in Martians, by the way.

CCN Do you think such life forms, primitive or not, actually exist somewhere in the universe?

SA I see no reason why not. In my view, and some would call it optimistic, there is no reason to suspect that intelligent life comes to an end or that some doomsday is awaiting us. But I'm a scientist, not a prophet. In general I favour a humanist world view of the kind advocated by the monist movement,[28] and I consider a universe with eternal life as part of this world view.

CCN Do you still subscribe to your panspermia hypothesis?

SA Oh yes, and the more I come to know about bacteria, the more confidence I have in its essential truth. For one thing, they are of the right size, actually corresponding to what Schwarzschild calculated[29] for the size of body that is most affected by the Sun's radiation pressure. How these panspermia could survive the intense cold during their journey through interplanetary and even interstellar space was a problem, but new experiments made in Kamerlingh Onnes' laboratory[30] prove that bacteria retain their germinating power for several years even under the most extreme cold in a vacuum. You look surprised, but that's how it is. It's actually an advantage, for the cold will slow down the life processes of the bacteria and in this way conserve their life for long periods of time. They are kind of hibernating in space.

So there is reason to have some confidence in the theory and further develop it. I am not worried that the bacterial spores will be killed by either starlight or cold, but I am a bit worried about the testability of my hypothesis, for at present I cannot see how it can be tested—and of what worth is a scientific hypothesis if it cannot be tested? It's a problem, but I am working on it.[31]

CCN It will be most interesting to see what you come up with.

SA Well, I think this makes a nice end for our conversation. Tomorrow will be another busy day with a meeting of the Nobel Committee, so

I'll say good night and return to my home. And I wish you a pleasant journey back to Germany. Rostock, you said?

CCN Yes, and then Berlin. Thank you, and good night professor. And thank you so much for your most interesting comments which have given me much to think about, so much food for thought. The world of science is amazing!

Notes

Svante August Arrhenius was born on 19 February 1859 at Vik, near Uppsala, Sweden, and he died on 2 October 1927 in Stockholm. After undergraduate studies at the University of Uppsala, he went to Stockholm in 1881 to prepare for his doctoral dissertation, an investigation into the electrical conductivity of electrolytes in which he proposed that an electrolyte (such as acetic acid) was ionized even in the absence of an external electrical field. The hypothesis was novel and controversial, which caused the examiners to grade his dissertation poorly. On the other hand, chemists elsewhere in Europe received it with sympathy and interest. In 1886 he went on an extended study tour abroad, which brought him in close contact with Wilhelm Ostwald and Jacobus van't Hoff, among others, and initiated the beginning of physical chemistry as a new and highly successful scientific discipline. Arrhenius' ionic theory of dissociation was controversial in some quarters, but in 1903 it earned him the Nobel Prize in Chemistry and 11 years later the Faraday Medal of the Chemical Society (see Figure 3). From 1900 until his death in 1927 he was an influential member of the Nobel Committee for Physics, using his power to favour scientists he liked and disfavour those he disliked.

In 1895 Arrhenius was appointed professor of physics at Stockholm College, the precursor of Stockholm University. Under his leadership the college became an important centre of "cosmical physics," a branch of science that largely consisted of meteorology, geophysics, astrophysics, oceanography, and physical chemistry. It was in this context that in 1896 he investigated how variations in the carbon dioxide content of the atmosphere cause climatic changes. Today this paper is considered a classic on the greenhouse effect. He also suggested a new theory of the aurora and of comets' tails based on the hypothesis of solar light pressure. Following his fruitful period at Stockholm College, in 1905 he became director of the new Nobel Institute for Physical Chemistry, a position he held until shortly before his death. In papers and books between 1901 and 1914 he developed a qualitative theory of the universe, which he insisted was infinite in space, matter, and time. *Worlds in the Making*, a book originally published in Swedish in 1907, attracted great public attention but little scientific recognition. At about the same time he made important contributions to immunology, which he wanted to provide with a chemical foundation. A versatile scientist who thrived on controversies, Arrhenius was still working until his death.

Biographical sources: Crawford (1996); Arrhenius et al. (2008); Riesenfeld (1931), which contains a bibliography of Arrhenius' writings.

1. *immunochemistry*: In 1907 Arrhenius gave a series of lectures at Berkeley, California, that were published as *Immunochemistry*. The main message of the book was that the chemical reactions involved in immunization follow the laws of physical chemistry. See Crawford (1996, pp. 167–226) for a discussion of Arrhenius' contributions to serology and immunology.

2. *a comprehensive textbook on cosmical physics*: Arrhenius' massive and encyclopaedic textbook of 1026 pages had chapters on a variety of subjects, including astrophysics, geophysics, oceanography, meteorology, atmospheric electricity, the aurora, and cosmogony (see Arrhenius 1903). A comprehensive survey of the interdisciplinary *fin-de-siècle* branch of science known as cosmical physics is given in Kragh (2013a). On Arrhenius as a cosmical physicist see Kragh (2013c).

3. *Austrian school of meteorology*: Cosmical physics obtained a firm institutional setting only in the Austro-Hungarian Empire, where it was cultivated in Vienna and Innsbruck in particular. To the Austrians, cosmical physics was primarily geophysics and meteorology, whereas astronomy and astrophysics only played a secondary role (see Kragh 2013a; Crawford 1996, pp. 132–134). Contrary to what Arrhenius intimates, not all members of the Austrian school excluded cosmological questions from their discussions of cosmical physics. Wilhelm Trabert, a professor at the University of Vienna, was a noteworthy exception.

4. *the hothouse effect*: In a paper of 1896 Arrhenius developed a theory of the ice ages based on calculations of how changes in atmospheric carbon dioxide influence the surface temperature of the Earth. He did not speak of a "greenhouse effect" and only referred *en passant* to the industrial emission of carbon dioxide. In a later work, Arrhenius (1908, p. 63) recognized the role played by "the enormous combustion of coal by our industrial establishments," but this he saw as an advantage and not a problem. "We may," he wrote, "hope to enjoy ages with more equable and better climates, especially as regards the colder regions of the earth, ages when the earth will bring forth much more abundant crops than at present, for the benefit of rapidly propagating mankind."

5. *he speculates that space may be curved*: It is conceivable, Trabert (1911, p. 259) wrote, that astronomical observations would one day demonstrate that "light, gravitation and electricity do not propagate in straight lines, but in circles." In sharp contrast to Arrhenius, but without referring to him, Trabert saw no problem in a materially finite universe developing towards a state of

maximum entropy. On Schwarzschild and the possibility of a curved-space universe, see Chapter 3. Arrhenius took infinite flat space to be self-evident, and he never referred to the possibility of a closed space.

6. *the hypothesis of energy quantization*: Max Planck proposed his celebrated hypothesis of discrete energy quanta in 1900, although initially without believing in their physical reality. At the time of the interview Arrhenius simply ignored quantum theory and the atomic theory proposed by Niels Bohr in 1913. This was not an unusual attitude at the time.

7. *I have a student, a Jew*: The student was Oskar Klein who joined Bohr in Copenhagen in early 1918; this was the beginning of a distinguished career in theoretical physics. The son of a German-born rabbi in Stockholm, Klein wrote his dissertation under Arrhenius on the theory of strong electrolytes. In 1930 he was appointed professor of physics at Stockholm University College. We shall meet him again in Chapter 11.

8. *a lecture Walther Nernst gave to the German Physical Society*: In an address of 28 January 1916 the eminent physical chemist Walther Nernst proposed that empty space is filled with electromagnetic radiation with a so-called zero-point energy, a concept of quantum theory that he carried over from material systems to the pure radiation field. He used the idea to argue for an eternal universe of the steady-state type, not unlike the one defended by Arrhenius. See Kragh (2012b) and also Chapter 5.

9. *we were once friends, but unfortunately*: Arrhenius first met Nernst in Würzburg in 1887, and the two young pioneers of physical chemistry became close friends. However, by 1900 the friendship had turned into a troubled relationship characterized by suspicion and bitter animosity. On their disagreements see Barkan (1999, pp. 217–238) and Coffey (2008, pp. 3–38). Arrhenius used his position in the Nobel Institute to discredit Nernst's scientific work and prevent him from being awarded a Nobel Prize in physics. In spite of Arrhenius' sustained opposition, Nernst was awarded the Nobel Prize in chemistry in 1921 (for the year 1920).

10. *your very successful book*: The hugely popular *Worlds in the Making* (Arrhenius 1908) was an English translation of *Människan inför Världsgåtan* published in Stockholm in 1907. It was translated also into German, French, Russian, Italian, Finnish, Czech, and Chinese and came in many later editions.

11. *Lebedev in Russia demonstrated it experimentally*: In a paper of 1901 the Russian physicist Peter Lebedev, working at Moscow State University, proved the existence of a light or radiation pressure in agreement with Maxwell's electromagnetic theory of light. In the same year the effect was independently confirmed by two American astrophysicists, Ernest Nichols and Gordon Hull at Dartmouth College. The idea that light exerts a pressure was already widely discussed in the eighteenth century, when it played an important role in connection with different theories of light, in particular the

Newtonian emission theory versus the wave theory favoured by Leonhard Euler and others. For a history of light pressure from Newton to Lebedev, see Worrall (1982). Arrhenius' paper of 1900, in which he used light pressure as a mechanism for cosmic processes, is dealt with in Kragh (2013c).

12. *What do we know about the gaseous nebulae?*: Arrhenius' view was not particularly unorthodox in the early years of the twentieth century. According to the Irish writer on astronomy Agnes Mary Clerke, the Orion nebula "contains inestimably less matter than should be comprised by it if its average density were that of a Crookes vacuum." Moreover: "On the supposition, however, that electrical discharges cause the glow of nebulæ, their average temperature . . . might approximate to absolute zero." (Clerke 1903, p. 522 and p. 535).

13. *the clockwork of the universe will never run down*: As stated in Arrhenius (1908, p. 195). For Arrhenius this was a premise, not a conclusion. For more on his cosmology, see Kragh (2013b). The clockwork metaphor for the mechanical universe is old, dating back to Nicole Oresme in the fourteenth century. It was later promoted by Leibniz, whereas Newton—with whose name the metaphor is often associated—never used it.

14. *exists eternally and, on a large scale, in roughly the same form*: His guiding principle was "the conviction that the Universe in its essence has always been what it is now. Matter, energy, and life have only varied as to shape and position in space." (quoted from Arrhenius 1908, p. xiv). Arrhenius thus anticipated what in the later steady-state theory of the universe would be called the "perfect cosmological principle." See Chapter 11 for more on this principle.

15. *those who are religiously motivated*: Arrhenius was a freethinker and atheist. An infinite universe was traditionally associated with materialist and atheist values, whereas Christian authors, whether Catholics or Protestants, typically argued for a finite universe. Still in 1916 the size of the universe was part of the cultural struggle, in Sweden and elsewhere, and Arrhenius' cosmology played a role in the struggle. See Kragh (2008a, pp. 172–173).

16. *Seeliger in Munich came up with an objection*: On Seeliger and his theoretical argument against an infinite stellar universe governed by Newton's law of gravitation, see Chapter 4. Arrhenius failed to understand the mathematical arguments leading to Seeliger's gravitational paradox.

17. *no reason to study his writings*: Arrhenius perceived Birkeland as a rival and seems to have been jealous of his widely acclaimed research on the aurora, a field Arrhenius considered himself a specialist in. He systematically refused to mention the name of his Norwegian colleague, pretending that he did not exist. He also used his influence to prevent Birkeland from receiving the Nobel Prize in Physics. See Chapter 1 and also Crawford (1996, p. 161).

18. *to quote from one of your papers*: The quotation (Arrhenius 1909, p. 229) is from a paper in the Italian multilingual journal *Scientia* founded in 1907.

It received wider attention after being translated into English in *The Monist* **21** (1911): 161–173.

19. *principles of the indestructability of energy and of matter*: Arrhenius operated with two independent conservation laws, of matter and of energy. According to relativity theory, the combined matter–energy quantity is conserved but not matter and classical energy separately. Surprisingly, Arrhenius never referred to Einstein's equivalence between mass and energy (as given by $E = mc^2$), although it was generally accepted by 1916.

20. *if the universe somehow came into existence*: Arrhenius' objection that an origin of the universe violates energy conservation was later repeated by Fred Hoyle, as mentioned in Chapter 11. However, it is a questionable argument: the law of conservation of energy implies that if a closed physical system at any given time $t^\star$ contains a total amount of energy, the same amount of energy must have been present in the system at $t < t^\star$. However, if the system is the entire universe, there is no time earlier than $t^\star = 0$.

21. *Clausius' law*: A reference to the German physicist Rudolf Clausius, who in 1850 established the second law of thermodynamics and in 1865 formulated it in terms of entropy, a name he coined. The second law was independently discovered by William Thomson, later Lord Kelvin, who preferred to speak of "dissipation" of heat or energy.

22. *a scientific law leads to a denial of science*: In a letter of 1907 to the American science writer Carl Snyder, Arrhenius admitted the premise of the so-called entropic argument, namely, the heat death scenario predicted by the law of entropy increase: "If there were an end, with complete rest, the condition would have been reached in the infinity of time which lies behind us." (quoted in Kragh 2013b, p. 19). Since the world has not yet come to an end, it seems to follow that the age of the universe must be finite. However, to Arrhenius this meant a created universe, which he considered a negation of science.

23. *something similar to Maxwell's demon*: In 1867 Maxwell construed a famous thought experiment in which fast-moving molecules are transferred from a colder to a hotter body, thereby increasing the entropy. However, Maxwell's experiment required the existence of a microscopic "demon" (or "finite being," as Maxwell called it) to open or close the valve between the cold and hot part of the system.

24. *Poincaré objected to what he called "Arrhenius' demon"*: In one of his last papers, Poincaré (1913, pp. 212–218) argued that Arrhenius' attempt to avoid heat death was unsuccessful. In his response to Poincaré, Arrhenius proposed a modification of his hypothesis. The letter is reproduced in Walter (2007, document 2.1).

25. *the recent edition of Newcomb–Engelmann*: This is a reference to the fifth edition of the widely read *Newcomb–Engelmanns Populäre Astronomie* (Leipzig: W. Engelmann, 1914). In a section written by Schwarzschild, he called

Arrhenius' mechanism a "daring and seductive hypothesis" (p. 732). On another occasion he proved that the temperature of the escaped molecules was less than the temperature of the nebula from which they escaped, contrary to Arrhenius' assertion. See Kragh (2013b) and also Chapter 3.

26. *I actually made the suggestion as early as 1901*: Arrhenius (1901, p. 872) argued that "the possibility of the existence of organized life forms has been secured for ever."

27. *I can't remember if I coined the name*: He did, but adopted it from the ancient Greek philosopher Anaxagoras. On Arrhenius' and others' ideas about life in space, see Dick (1996) and Kamminga (1982).

28. *a humanist world view of the kind advocated by the monist movement*: Together with his friend, the chemist Wilhelm Ostwald, and other leading monists, Arrhenius participated in the first International Monist Congress held in Hamburg in 1911. The German Association of Monism, of which Arrhenius was a member, was founded in 1906 by the evolutionary biologist and philosopher Ernst Haeckel to promote a scientific, progressive, and atheistic world view. The monist movement was popular in scientific and philosophical circles before World War II.

29. *corresponding to what Schwarzschild calculated*: In a paper of 1901 Schwarzschild revised Arrhenius' theory of light pressure by taking into considerations diffraction effects. His calculations showed that the bodies most affected by the Sun's radiation must have a diameter of 0.00016 mm, of the same order of magnitude that Arrhenius estimated for bacteria (see Arrhenius 1908, pp. 97–99, 220).

30. *experiments made in Kamerlingh Onnes' laboratory*: The Dutch physicist Heike Kamerlingh Onnes, a Nobel laureate in 1913 for his discovery of liquid helium, was director of a famous cryogenic laboratory in Leiden. In 1911 he discovered superconductivity, the surprising vanishing of electrical resistance at very low temperature.

31. *but I am working on it*: Arrhenius' last scientific paper, published posthumously in 1927, dealt with thermophilic bacteria and was an attempt to provide empirical support for the hypothesis of panspermia.

3

Karl Schwarzschild: Astronomer and Physicist

Interview conducted at Schwarzschild's home in Potsdam outside Berlin on 19 March 1916. Language: German.

Had it not been for Ejnar Hertzsprung I would not have had the opportunity to interview the eminent astronomer Karl Schwarzschild about his views on the universe. While in Copenhagen as a student of applied chemistry I worked on and off for a time as an unpaid assistant at the private Urania Observatory, and there I met Hertzsprung, who at the time had already made important contributions to astronomy. He found it interesting that I was studying chemical engineering at the Polytechnic College, the same institution from which he had graduated as a chemical engineer about a decade earlier. Later, when we both moved to Germany, we met a couple of times, talking about astronomy and sharing gossip about people we knew in Copenhagen. In 1909 Hertzsprung had been hired by Schwarzschild to a position at Göttingen University, and later the same year he followed Schwarzschild to Potsdam. This was shortly after his first version of the famous diagram—today the Hertzsprung–Russell or H–R diagram—with which his name is associated.

My original plan to interview Schwarzschild came to nothing as I was unable to meet him. It turned out that at the time he was in the artillery staff on the Eastern Front, taking part in the Russian campaign. I thought of postponing the interview until the war had ended, but then I was told that he had contracted a serious skin disease called pemphigus, known to be fatal. He was in the process of being sent back to Potsdam as an invalid. At first I felt it improper to approach him under these circumstances, but when I aired my moral scruples to Hertzsprung he offered to bring up the subject with his friend, assuring me that he was still clear in his mind and scientifically active. And thus it happened that I met

Figure 4 Karl Schwarschild. Source: Physikalische Zeitschrift 17 (1916): 545.

Schwarzschild (Figure 4) in his home in the early spring of 1916, shortly after my interview with Arrhenius in Stockholm. Schwarzschild passed away less than two months later.

CCN I greatly appreciate your willingness to talk with me, and I promise you that the interview will be short. If you feel tired or suffer pain, please let me know and we will take a break or just terminate the interview. We shall not talk about your illness, but I imagine it makes it difficult for you to work and concentrate on matters of science?

KS Strangely, in spite of my illness—and it *is* painful—I feel in great intellectual shape and have lately done some interesting work which hasn't been published yet. I thank the merciful God that my capacity for work has not diminished, for if I couldn't focus on science the present situation would be unbearable. My condition may worsen, but so far I can still think and calculate, and that's a great relief. But no, we shall not talk of my dreadful situation. . . . Hertzsprung tells me that you are interested in cosmology?

CCN Yes, very much so. Perhaps I could start by asking if you consider yourself a cosmologist?

KS A cosmologist? That's a funny question, not to mention that it's a funny word. Of course I do not, I'm a physicist and an astronomer, and perhaps an applied mathematician as well. I have dealt from time to time with questions of a cosmological nature, but that doesn't make me a "cosmologist." I know there are people who think of themselves as cosmologists, but they are mostly philosophers or amateur scientists—and, believe me, more often cranks than not. I remember reading a few books by self-styled cosmologists when I was in my teens. There was a book by some author[1]—Sonnenschmidt, I think his name was—which fascinated me a great deal at the time but which I also realized was purely speculative. It was a sobering experience. No, in my view cosmology is simply that branch of physics and astronomy which deals with the universe in its entirety. It's no more than that, but then that's enough, isn't it?

CCN It certainly is. Well, you are one of the most distinguished theoretical astronomers of our time, known in particular for work you did while a professor in Göttingen and subsequently in Potsdam. But I would like to pay attention to one of your earlier contributions that you may not yourself consider very important and which is not, in fact, well known among astronomers. I am referring to a lecture you gave to the Astronomische Gesellschaft during its meeting in Heidelberg in 1900, at which time you were working in Munich, I think.

KS Ah, that paper, the one on the curvature of space.[2] Yes, I guess you are right, it didn't make much of an impact and I didn't myself think very highly of it, neither at the time nor now. It was a kind of survey addressed to a general audience of astronomers, not to mathematicians. But it was an interesting problem that had not been previously investigated and I enjoyed working it out. It was later extended by Harzer, my colleague in Kiel,[3] who investigated models of the Milky Way universe—Seeliger's models, to be precise—in a space of constant positive curvature. He found that it would take 10,000 light-years or thereabouts for a ray of light to circumnavigate the closed stellar universe. This is interesting, but neither Harzer nor I consider it more than a possibility. As yet we have no evidence that cosmic space is *really* finite and with a positive curvature. One should take the possibility seriously, but so far without being committed to it—keep an open mind, you know.

CCN Yes, that appears to be a sound advice.

KS Anyway, although non-Euclidean geometry was far from a novelty in 1900[4] it was considered a field of pure mathematics, whereas astronomers either ignored it—as most did—or, in a few cases, applied it to local astronomical phenomena, such as the problem of the motion of Mercury's perihelion. My idea was different, it was to look at cosmological space, so to speak, the space in which everything material is located. Is this global space curved? Might it be closed, and the universe therefore finite? Or, for that matter, might its geometry be hyperbolic? These were the kinds of questions I addressed, but of course without coming up with a definite answer. There still is no definite answer, not even a likely one.

CCN Incidentally, did you know that as early as 1872 Zöllner in Leipzig suggested a closed model of the universe,[5] using Riemann's geometry?

KS No, I didn't know that, I wasn't even born then. I have been told stories about this Zöllner, possibly by Seeliger, and I'm aware of his work in stellar photometry. But I haven't read him. Did he really propose a closed model?

CCN Yes, I looked his book up in the library. He thought that the assumption of a positively curved space might solve Olbers' paradox and other problems of the infinite universe, but his arguments are hard to follow because they are such an obscure mix of physics and philosophy. I don't think they made any impact on astronomers at the time.

KS I certainly didn't know about them. As far as I recall I came to my ideas independently.

CCN No doubt you did. So what was the overall result of your paper?

KS Well, let me try summarizing it. By using the best astronomical data available at the time—concerning star counts and stellar parallaxes—I simply put observational bounds on the curvature of space, whether positive or negative. If my memory doesn't fail me, then . . . for a hyperbolic space I concluded that the radius of curvature[6] had to be greater than at least a million astronomical units, which makes it observationally indistinguishable from flat space.[7] In the case of a closed universe, a so-called elliptic space, I made certain assumptions that I later came to recognize as unrealistic, and I came up with a value of about 100 million astronomical units, I think, but I am not sure. . . . Oh, I see that you brought a copy of my paper with you, is my memory correct?

CCN   By and large, I can see that the figures in your paper are 4 million and 160 million, respectively, both in astronomical units. Now, as Zöllner was the first to realize, the problem of curved space relates to the old question of whether the universe is finite or not. This is of course a question that has been much discussed and I believe your former professor has an opinion about it. Have you formed an opinion?

KS   Not really, and certainly not a strong one. It's as much a philosophical as a scientific question. But I do not quite agree with Seeliger[8] and tend to consider the possibility of closed space attractive—just that, attractive. It's a kind of prejudice, you know, an opinion from a philosophical and human perspective, or perhaps it's based on my view of the theory of knowledge, I don't know. It's just that personally I don't feel at home in an infinite universe, if you understand what I mean. It sort of frightens me, not so much emotionally as intellectually. On the other hand, I know that there are people who think differently, Arrhenius for example. I wonder if I said something about that in my old paper?

CCN   You actually did, let me check where it is. Hmm . . . , you say that if the stellar universe is closed and finite, then—and now I quote— "a time will come when space will be investigated like the surface of the Earth, where macroscopic investigations are complete and only the microscopic ones need continue." And then you proceed: "A major part of my interest in the hypothesis of an elliptic space derives from this far reaching view." That's it.

KS   Right, now I remember.

CCN   You mentioned Arrhenius, with whom I actually had a conversation just a few weeks ago in Stockholm. He is not only convinced that the universe is infinite but equally convinced that it must be in a perpetual steady state—I mean, *really* convinced. He just doesn't believe in a gravitational collapse or in the thermodynamic heat death, or in anything at all that can shatter the overall peace of the universe and lead to a cosmological Armageddon—or birth.

Have you considered the cosmic history of the universe or contemplated a possible beginning of it?

KS In private I have, but only in private. Frankly, I don't have much confidence in Arrhenius' reasoning, which seems to rest on a qualitative basis and, as you point out, on the a priori assumption that the universe is infinite and eternal. Perhaps it is, but how can we possibly know?

I once wrote a review essay on Poincaré's book on cosmogony,[9] in which I sided with his criticism of Arrhenius' attempt to turn the nebulae into counter-entropic machines and thereby avoid the heat death. In fact, I proved by means of a simple example based on the kinetic theory of gases that the hypothesis doesn't work. For the time being we must rest content that we simply have no reliable knowledge concerning the temporal development of the universe. We can speculate about it, but that's all. Perhaps the principle of simplicity speaks in favour of a static universe. At least, it's difficult to comprehend a beginning of it. On the other hand . . . , no, I really don't know.

CCN Arrhenius is also well known for his arguments that life is abundant in the universe. One reason for the popularity of his books is undoubtedly the public fascination with the possibility of intelligent life elsewhere in the universe. This is a question that evidently invites speculation, but it's nonetheless a natural one, at least to my mind. I just wondered if you have any opinion on this matter?

KS You're right, it's a big business in popular astronomy and has always been. I have often been asked about it and would wish that I had an answer. There certainly is no reason to take Martians seriously or, for that matter, Arrhenius' speculations, but on a much larger cosmic scale one can perhaps come up with probability arguments that there are civilizations out there[10] and that some of them are even more advanced than ours. For example, the number of civilizations may be estimated by the number of stars multiplied by the probability that they have Earth-like planets and the probability that life develops on the planets . . . and so on. Unfortunately we have no idea of these individual probabilities, so I'm just speculating.[11] And yet I don't think that the possibility of other civilizations should just be dismissed out of hand. On the other hand, if they exist we most likely will never know, except that there is the remote possibility that one day we may receive radio signals . . . no, I'm daydreaming.

CCN Thanks for sharing your dreams with me, professor. This was really a digression, so please continue with what you were about to say before I interrupted you.

KS Let me see, I think I was going to say that what we are missing in forming a picture of the universe as a whole are new data and some new theoretical perspective, and most likely this is also the case when it comes to the so-called history of the universe. Let me just mention that Slipher, an American astronomer at Lowell's observatory,[12] reported some years ago that the light from several spiral nebulae is shifted towards the red end of the spectrum, indicating that they are moving away from the solar system. I once met him at a meeting[13] on solar physics in California, but that was before he had measured these rather mysterious redshifts, as they are called.

Unfortunately I'm not up to date, you know that the war makes scientific communication with foreign nations so difficult. Although the Americans are not part of the war,[14] at present we have no access to their publications, which is most unfortunate. If the redshifts are confirmed and turn out to be systematic, they may be important. They may even be of some cosmological significance, although at present I cannot see clearly in what way. But Hertzsprung is quite impressed,[15] you know, he thinks that the data prove that the spiral nebulae do *not* belong to the Milky Way system. And Eddington in England also believes the nebular redshifts are of great importance. I wonder what their true meaning is.

CCN You mentioned the need for a new theoretical perspective. Might this come from Einstein's very new theory of gravitation? I understand that you, perhaps alone among German astronomers, have taken an interest in Einstein's theory and even contributed to it. I have only read about it and, frankly, I understand little of it. It's beyond my reach, I'm afraid.

KS Oh yes, I have taken an interest in it, even a burning interest. But we should not forget Freundlich,[16] who studied under me in Göttingen, for he has also examined the astronomical consequences of the theory. But then we may be the only ones so far. I don't think Eddington knows about it.

CCN Because of the war?

KS Yes, because of the war. In fact, I became interested in Einstein's gravitation theory at an early stage of its development, trying—unsuccessfully I have to admit—to verify in the solar spectrum the gravitational redshift that Einstein predicted . . .[17]

CCN Sorry, you just mentioned the redshifts observed by this American astronomer, is that what Einstein has predicted?

KS No, no, whatever the cause of the spiral redshifts is, it is something entirely different, it's much too large to be gravitational and it seems to be restricted to the spiral nebulae.

CCN Oh, I see. So there are two kinds of redshifts?

KS Yes, and they are quite different. Well, quite recently, after Einstein and Hilbert had obtained the right equations,[18] I wrote a couple of mathematical papers in order to clarify what the equations are all about. I did this work while serving at the Russian Front, you know, but fortunately I had enough spare time in between my calculations of artillery trajectories, and I received Einstein's latest communications to the Prussian Academy with almost no delay. Being absent from Berlin I sent my papers to Einstein, who was kind enough to present them to the Academy, the first in January and the second in February. So it's very new.

CCN You have indeed been busy.

KS Yes, and that's not all. Another new theory I have become interested in, if I can briefly change the subject, is the quantum theory of atoms, which has been some consolation for me during my illness. I recently corrected the proofs of a paper that will appear in the proceedings of the Berlin Academy, it's on the electric splitting of spectral lines[19] according to the theory of Bohr and Sommerfeld. It relies on mathematical methods from celestial mechanics that are well known to astronomers but not to physicists, and I was fascinated to see how these powerful methods are also applicable to the tiny world of atoms. It confirms my belief in the unity of science.[20] But this was just a side-remark—it hasn't been presented to the Academy yet, and unfortunately I am unable to do it myself.

CCN I'm deeply impressed! Is it possible to explain in non-mathematical language what your papers on Einstein's theory[21] are about?

KS No, not really, the mathematics is all-important, but let me try. Hmm, I succeeded in finding the first complete *and* unique solutions to the field equations in the case of a spherically symmetric gravitational field surrounding a mass point, and in this way I came to an exact solution of the Mercury problem[22] instead of the approximate result Einstein had found.

CCN This is the old problem of Mercury's elliptical orbit rotating slowly at a rate that doesn't agree with classical mechanics?

KS Yes it is, and Einstein brilliantly solved it, demonstrating that in the first approximation general relativity gives Newton's law and in the second approximation it reproduces the known anomaly in Mercury's motion. But he didn't go further.

And then I also used the theory, and this time the fully covariant Einstein equations, to examine the gravitational field of an idealized fluid sphere—something like a star, but not really a star for it's more like a mathematical toy model of a star. At any rate, what I found was astonishing. If the volume of the sphere diminishes and the density thus continues to grow, as measured by an observer outside the sphere, there will be a limit at which the equations make no physical sense. There will occur what I call an external singularity corresponding to a space curvature so great that even light cannot escape the compressed sphere. Do you follow me?

CCN Not really, I need time to swallow it, if I can swallow it at all. But just go on, please. It may beyond my reach, but it's fascinating nonetheless.

KS Whether this scenario I have outlined has a physical significance or not I cannot say, but the bodies have to be smaller than their gravitational radius,[23] which in the case of the Sun means that it has to be squeezed down to the size of 3 kilometres. Think about the density of such a body! For a mass of 1 gram the size is ridiculously small, much smaller than an atomic nucleus, 10^{-28} cm I think—and the density is even more ridiculously high. It's all very weird and perhaps just a mathematical curiosity. On the other hand, history tells us that the mathematical solutions are often realized in nature, as if there were some kind of pre-established harmony between mathematics and physics.[24] This idea is quite strong in Göttingen, and I am not foreign to it myself. You may even call me a believer.

CCN Weird it is, fascinatingly weird. If it relates to nature at all one could perhaps imagine, just as a speculation, some as yet undiscovered kinds of extremely massive stars. . . . Could one also imagine that your analysis relates to the universe as a whole?

KS No, I don't think so. It has nothing to do with cosmology. After all, so far there doesn't even exist a cosmological theory based on Einstein's field equations.[25] On the other hand, about a month ago, when I was still at the Russian Front, I wrote Einstein a letter in which I pointed out that the equations have a solution corresponding to a closed universe with an elliptic geometry—actually of the same kind that I considered in my old paper of 1900 that we discussed. I don't know if he's working on it, perhaps he's too busy with other things.

CCN These new ideas of Einstein that you mention, are you saying that they may give a kind of physical justification of a closed universe?

KS Well, yes, perhaps, although it's too early to say. But it's an interesting thought, for I used to think that the closed universe was no more than a speculation, and now . . . who knows? Some years ago I wrote a popular article on cosmogony in which I dismissed the idea as unfounded[26]—it was in the Newcomb–Engelmann book—but thanks to Einstein I'm not so sure any longer. . . . Oh, you said that we could break up the interview if I got tired . . . I must admit that . . .

CCN Absolutely, we'll stop right now. I have already taken up too much of your time. Thank you so much. I should send you the heartiest regards from Hertzsprung.

Notes

Karl Schwarzschild was born on 9 October 1873 in Frankfurt am Main and died on 11 May 1916 in Potsdam. A prodigy, he wrote his first scientific papers, dealing with the theory of the motion of double stars, while a 16-year-old student at Frankfurt Gymnasium. Having completed high school he studied astronomy in Munich, where he completed his doctoral dissertation under the supervision of Hugo von Seeliger, whose teaching had a lasting influence on him. He subsequently obtained a position at the Von Kuffner Observatory near Vienna, and after a brief stay in Munich he was appointed director of Göttingen Observatory and also, in 1902, to a full professorship. During his period in Göttingen (from 1901 to 1909) he collaborated with famous physicists and mathematicians such as David Hilbert and Hermann Minkowski. His scientific work was wide-ranging, including topics such as the electron theory of matter, radiation equilibrium in the solar atmosphere, the effect of light pressure from the Sun, and a large survey of stellar magnitudes that led him to a model of the Milky Way universe. Whether in astronomy, physics, or astrophysics, much of Schwarzschild's work was characterized by the use of advanced mathematics. Arthur Eddington (1917, p. 319) was impressed by his amazing versatility: "His joy was to range unrestricted over the pastures of knowledge, and, like a guerrilla leader, his attacks fell where they were least expected."

At the end of 1909 Schwarzschild left Göttingen to take up a position as director of the Astrophysical Observatory in Potsdam. He brought his assistant, the Danish astronomer Ejnar Hertzsprung, with him. In 1912 he was elected a member of the prestigious Prussian Academy of Science. Physics was at the time in a state of transition, and Schwarzschild eagerly took up the new and exciting fields of quantum theory and the relativistic theory of gravitation. During the final year of his life, he made important contributions to the Bohr–Sommerfeld atomic theory and Einstein's theory of gravitation. His work on the latter subject later became recognized as the foundation of the physics of black holes. With the outbreak of World War I, Schwarzschild volunteered for military service. In 1915, when he was serving as a member of the artillery staff at the Russian Front, he contracted a rare and at the time incurable skin disease affecting the immune system. He was invalided home in March 1916. After a stay in a hospital, he died in May 1916.

Biographical sources: Eddington (1917); Hertzsprung (1917); Dieke (1975); Schwarzschild (1992) is a collection of his works.

1. *a book by some author*: The book that Schwarzschild read as a teenager was Hermann Sonnenschmidt's *Kosmologie: Geschichte und Entwicklung des Weltbaues* (Cologne: E. H. Mayer, 1880), a time-typical example of philosophical cosmology in the positivist and materialist tradition that was popular in Germany at the end of the nineteenth century.

2. *that paper, the one on the curvature of space*: This paper (Schwarzschild 1900) has been translated into English in *Classical and Quantum Gravity* **15** (1998): 2539–2544. Schwarzschild gave his address to the Astronomische Gesellschaft [Astronomical Society] on 9 August 1900. For background and analysis see Schemmel (2005) and Kragh (2012c).

3. *extended by Harzer, my colleague in Kiel*: In 1908 the German astronomer Paul Harzer, a professor of astronomy at the University of Kiel, published a paper on the stellar system in a space of constant positive curvature. His model universe consisted of a stellar system (the Milky Way) surrounded by a closed cosmic space with a volume about 17 times that of the material universe. For Harzer's paper and its relation to Schwarzschild's earlier paper, see Kragh (2012c).

4. *non-Euclidean geometry was far from a novelty in 1900*: The insight that there exist geometries different from Euclid's dates from about 1830, when hyperbolic space was investigated by the Russian mathematician and astronomer Nicolai Lobachevsky. During the second half of the nineteenth century more than 3000 articles and books were published on the subject, but the general view was that real space is flat or Euclidean. Only a handful of astronomers and physicists considered the possibility that it was not.

5. *suggested a closed model of the universe*: The astrophysicist Carl Friedrich Zöllner published his ideas in 1872 in a book entitled *Über die Natur der Cometen*. As he pointed out on p. 308, "The assumption of a positive value of the spatial curvature measure involves us in no way in contradictions with the phenomena of the experienced world if only its value is taken to be sufficiently small." In other publications from the 1870s he defended the unorthodox idea of a four-dimensional space including a spiritual fourth dimension. For details, see Kragh (2012a).

6. *radius of curvature*: A measure of the constant space curvature. The greater the radius of curvature is, the less curved space is. An example is the two-dimensional surface of a sphere with varying radius; for a very large radius the surface becomes almost flat. The radius of curvature R relates to the curvature K and the curvature parameter k by $R^2 = k/K$, where k can attain the values 0 (Euclidean space), +1 (spherical or elliptic space) or −1 (hyperbolic space).

7. *observationally indistinguishable from flat space*: Schwarzschild (1900, p. 342) concluded: "Thus the curvature of the hyperbolic space is so insignificant that it cannot be observed by measurements in the planetary system, and because hyperbolic space is infinite, like Euclidean space, no unusual appearances will be observed on looking at the system of fixed stars." The "elliptic space" that Schwarzschild talks about is not quite the same as what usually is called spherical space. While in spherical or Riemannian space all geodesics from a given point intersect again in an "antipodal point" at a distance πR, in elliptic space two geodesics have only one point in common. Both spaces are finite and with constant curvature, but for the same radius of curvature the volumes differ. For spherical space the volume is $2\pi^2 R^3$, and for elliptic space it is $\pi^2 R^3$. The name "elliptic" is unfortunate since it suggests a direct connection with an ellipse or visualization of this kind of space in two dimensions as the surface of an ellipsoid.

8. *I do not quite agree with Seeliger*: According to Seeliger, curved space was a meaningless concept. His views on space and the universe are covered in the interview in Chapter 4.

9. *a review essay of Poincaré's book on cosmogony*: Schwarzschild's review of Henri Poincaré's *Leçons sur les Hypothèses Cosmogoniques* (Paris: A. Hermann et Fils, 1911) appeared in *Astrophysical Journal* **37** (1913): 294–298.

10. *arguments that there are civilizations out there*: On this question, Schwarzschild suggested: "Most likely, there are only relatively few planets populated with rational beings. However, if we contemplate that there may be hundreds of millions of planets, then even a small fraction may really represent a considerable number, and many of them may be populated by beings greatly superseding humans in intellectual powers." See Kemp (1911, p. 675).

11. *I'm just speculating*: The kind of argument Schwarzschild had in mind would much later be formalized in the so-called Drake equation named after Frank Drake, an American radioastronomer. Drake's equation of 1961 gives the number of alien civilizations in terms of the product of various probabilities of life and of communication.

12. *an American astronomer at Lowell's observatory*: In 1912 Vesto Melvin Slipher at the Lowell Observatory in Flagstaff, Arizona found that the spectral lines of the Andromeda nebula were shifted towards the blue end of the spectrum. He soon realized that it was an exception and that the spectra of spiral nebulae are generally shifted towards the red. If interpreted as a Doppler effect, some of the measured redshifts indicated recessional velocities of more than 1000 km/s. For details on Slipher's life and work see Way and Hunter (2013). The nebular redshifts were only seen as cosmologically significant in the 1920s, and by about 1930 they were understood as a result of the expansion of the universe. The interpretation was first suggested by Lemaître in 1927, as discussed in Chapter 7. Although Slipher's redshift measurements were of crucial importance, it is absurd to claim, as Hoyle (1994, p. 277) did, that "it was Slipher who discovered the expansion of the Universe."

13. *I once met him at a meeting*: See the picture in Berendzen et al. (1976, p. 79), showing Slipher, Schwarzschild, and other astronomers at a conference on solar research held in September 1910 at Mount Wilson, California.

14. *the Americans are not part of the war*: The United States remained formally neutral in World War I until 6 April 1917, when the country declared war on Germany.

15. *but Hertzsprung is quite impressed*: In a letter to Slipher of 14 March 1914, Hertzsprung wrote: "It seems to me, that with this discovery the great question, if the spirals belong to the system of the Milky Way or not, is answered with great certainty to the end, that they do not." (quoted in Smith 1982, p. 22). Slipher was more cautious, and only in 1917 did he agree with Hertzsprung that the redshift observations favoured the so-called island universe theory.

16. *we should not forget Freundlich*: A reference to the German physicist and astronomer Erwin Freundlich, who from an early date worked to verify the astronomical predictions based on the new theory of general relativity. See also Chapter 4.

17. *the gravitational redshift that Einstein predicted*: Einstein had deduced the redshift due to gravitation as early as 1911, as a consequence of the principle of equivalence. If light is emitted at wavelength λ from the surface of a spherical mass M of radius R, it will be received on Earth at wavelength $\lambda + \Delta\lambda$, where $\Delta\lambda/\lambda = GM/c^2R$. The same result came out of the covariant field equations of gravitation that he derived in November 1915.

18. *Einstein and Hilbert had obtained the right equations*: The famous mathematician David Hilbert presented the basic equations of general relativity to the Prussian Academy on 20 November 1915, five days before Einstein. Nonetheless, it was Einstein alone who discovered general relativity. For the complex relationship between the works of Hilbert and Einstein, see, for example, Rowe (2001).

19. *the electric splitting of spectral lines*: In late 1913 the German physicist Johannes Stark discovered that in the presence of a strong electric field the spectral lines of hydrogen separate into components. The Stark effect immediately became a testing ground for Bohr's new atomic theory and also attracted interest from astrophysicists. A complete solution within the framework of the old quantum theory was provided by the theory of Schwarzschild and independently by the physicist Paul Epstein. Schwarzschild's paper was presented to the Prussian Academy in Berlin on 30 March 1916.

20. *my belief in the unity of science*: In his speech of admission to the Prussian Academy of Science in 1913, Schwarzschild said: "Mathematics, physics, chemistry, astronomy, march in one front. Whichever lags behind is drawn after. Whichever hastens ahead helps the others. The closest solidarity exists between astronomy and the whole circle of exact science." (quoted in Eddington 1917, p. 319).

21. *your papers on Einstein's theory*: Einstein presented the two papers to the Prussian Academy in January and February 1916. They have been translated into English, the first in *General Relativity and Gravitation* 35 (2003): 951–959 and the second in *Arxiv*: 9912.033 [physics. hist-ph] (1999). Although these papers are often described today as the foundation of the physics of black holes, what Schwarzschild dealt with was not really black holes in the later meaning of the term. His papers were of a mathematical nature and he had little to say about the physical meaning of the "Schwarzschild singularity" appearing in his formulae. Only in the late 1930s did it become clear that, in the case of a collapsing massive star, Schwarzschild's singularity defines a surface from which light cannot escape, what today is known as the "Schwarzschild horizon" for a non-rotating black hole. On the early interpretations of his work, see Eisenstaedt (1989).

22. *the Mercury problem*: This problem or anomaly dates from 1859, when it was recognized that Mercury does not move around the Sun exactly as predicted by Newtonian mechanics. The speed of rotation of the planet's perihelion

differed from the predicted value by approximately 43″ per century. The problem persisted until 18 November 1915, when Einstein presented his calculation of a value that fully agreed with observations. For a comprehensive history of the anomaly, see Roseveare (1982).

23. *bodies have to be smaller than their gravitational radius*: The gravitational radius, or Schwarzschild radius, is given by $R = 2GM/c^2$, where G is the gravitational constant, M is the mass of the body, and c is the velocity of light in a vacuum. Numerically, $R(\text{m}) = 1.5 \times 10^{-27} \times M$ (kg).

24. *some kind of pre-established harmony between mathematics and physics*: Many German mathematicians and theoretical physicists in the early part of the twentieth century were convinced of a close harmony between mathematical and physical truths. The belief was particularly strong in Göttingen, where it was espoused by Hilbert and Minkowski, among others. In their version, the harmony implied that mathematics was the royal road to progress in physics and astronomy, or even that the physical sciences could be reduced to pure mathematics (which was not Schwarzschild's view). See Pyenson (1982) and Kragh (2011, pp. 79–83).

25. *a cosmological theory based on Einstein's field equations*: By March 1916 Einstein had completed his general theory of relativity but not yet applied it to the universe. In a letter of 6 February 1916, Schwarzschild wrote to him: "As concerns very large spaces, your theory has a similar position as Riemann's geometry, and you are certainly not unaware that one obtains an elliptic geometry from your theory if one puts the entire universe under uniform pressure." (Einstein 1998, document 188). Einstein presented his closed cosmological model governed by the new gravitation theory, the beginning of relativistic cosmology, to the Prussian Academy on 8 February 1917.

26. *I dismissed the idea as unfounded*: In a contribution on cosmogony to the fourth edition of *Newcomb–Engelmanns Populäre Astronomie* (Kemp 1911, p. 664), Schwarzschild called the hypothesis of a closed Riemannian universe "much too speculative to be [further] discussed." In the same article he criticized Arrhenius' cosmological ideas as outlined in Chapter 2.

4

Hugo von Seeliger and Stellar Cosmology

Interview conducted in Seeliger's office at the Munich Observatory in Bogenhausen, Munich, on 20 November 1920. Language: German.

Apart from my visits to Copenhagen and Stockholm in the winter of 1916, I stayed in Germany for most of the war. The only other exception was a week's vacation in Switzerland. I wanted to do an interview in Munich with the astronomer Hugo von Seeliger, whom I considered an eminent representative of classical stellar cosmology. However, I had to postpone the meeting to the autumn of 1918, and then it was no longer possible because of the tumultuous situation in Germany following the military defeat. In fact, to avoid the turmoil I preferred to return to peaceful Denmark, where I stayed until the following spring.

By 1920 the situation had improved somewhat and Seeliger agreed to meet with me. His health was failing and he had wanted to retire when he turned seventy, but forced by the economic and political circumstances he had remained in his post. His original scepticism with regard to the interview waned when he understood that I had already interviewed Schwarzschild and also that I was acquainted with Hertzsprung (who in 1919 had left Potsdam for Leiden). He was pleased to point out that Hertzsprung had married a daughter of his esteemed colleague in the Netherlands, Jacobus Kapteyn. In spite of Seeliger having a reputation for being a difficult person, he received me in a friendly manner and the interview went smoothly. I had prepared it in advance and discussed with him a few of the questions that I wanted to ask.

CCN Professor Seeliger [Figure 5], thank you very much for taking the time to share with me some of your thoughts about the structure of the universe. You are a distinguished veteran astronomer who has been investigating the mysteries of the heavens for a long time; for

Phot.1909.

Figure 5 Hugo von Seeliger. Source: Kienle 1925, p. 1.

the last couple of decades you have focused on the distribution of stars in the Milky Way system. I understand that this great work has recently been completed. So, before we proceed to other matters, perhaps you could tell me something about your contributions to what is sometimes called statistical cosmology?[1]

HvS Well, yes, it is correct that in March this year I presented to the Bavarian Academy of Sciences—of which, by the way, I have had the honour to serve as president since 1918—a comprehensive summary account of work that has gone on for more than two decades. Had it not been for the terrible war I would have presented it earlier. Anyway, it is published in a lengthy memoir of the proceedings of the academy under the title *Untersuchungen über das Sternsystem* [Investigations on the Stellar System], for this is what it is about—it's a careful investigation of the stellar system.

You see, the problem which has occupied me and other astronomers for quite a long time is how the number of stars—or the star density—varies in space as seen from our location in the solar system. This is an extremely complicated problem that requires not only good observational data but also, and this is no less important, advanced mathematical methods to make sense of the data. My colleague Professor Kapteyn in Groningen[2] has worked on the same problem, but I believe that I was the first to develop the mathematical theory of the statistics of stellar distributions. I did that by using integral equations—are you familiar with this tool?

CCN Well, no, I must admit that I'm not. Perhaps we could skip the higher mathematics and you could just tell me about the results of this interesting line of reasoning. Would that be agreeable to you?

HvS Of course, you are not a mathematician, so I shall avoid the mathematical arguments, although integral equations are really essential and very useful. But I need to say that some of the mathematical arguments rely on improvements made by Schwarzschild,[3] who, as you know, was a former student of mine but now unfortunately is no longer with us. It is so tragic, and such a great loss to German science, that he died while in the prime of his life. I was quite moved when you told me of how you had this conversation with him shortly before he passed away . . . it's such a sad story. Now, let me see, you asked me about the overall picture I have formed of the stellar universe?

CCN Yes, please.

HvS First of all, I have come to the conclusion that the stellar system— or the Milky Way if you prefer that name—has a kind of ellipsoidal shape with the Sun close to its centre but not right at it. As to the dimensions of the system, my best estimate is that its maximum extension in the galactic plane is about 33,000 light-years and some 3900 light-years in the direction toward the galactic poles. I would like to emphasize that a stellar universe of roughly this kind and size agrees, at least approximately, with the conclusion that Schwarzschild arrived at and also with Kapteyn's recent picture[4] of the universe based on methods different from mine. So there is little doubt that in its main outline it is correct.

CCN I read in a magazine that recently the problem of the size and structure of the Milky Way has been the subject of a public discussion in the United States. Professor Shapley of the Mount Wilson Observatory was one of the discussants, defending his view of a monster galaxy, whereas the more traditional view was argued by . . . no, right now his name escapes me. Do you know about this great debate on

the universe?[5] Although I don't know the details I find such a dialogue interesting, there's something Galilean about it.[6] Oh, now it
pops up, the name of Shapley's opponent, it's Curtis. Yes, Curtis.

HvS   I have been told about this so-called great debate, but I don't like it
at all. Questions concerning the stellar universe should be discussed
in scientific meetings and communications and not in the public
arena, as if it were a music hall performance. The view with the most
applause is the correct one! Ha! But that's typical of the American
style, where public visibility, promotion, and media appeal count
as much as strictly scientific arguments. Of course I know about
Shapley's picture of the galactic system,[7] but I consider it to be plain
wrong. It's just out of the question that the Milky Way has such exces-
sive dimensions as he claims, and it's also wrong that the Sun is very
far from its centre. I suspect it's just another example of American
sensationalism and tendency to megalomania.

CCN   I see, that's an interesting perspective. According to your own
view, the multitude of stars is spread over the vast ellipsoid you mentioned. I wonder, what is outside it? And another question: is the
density of stars the same throughout the system?

HvS   No, that is far from the case, for the number of stars in space
diminishes with the distance from the Sun and does so in a particular
way revealed by my analysis. One might believe that the star density
fades away continuously, but my analysis actually indicates that the
density, after diminishing for a great distance, drops nearly abruptly
to zero. If this should turn out to be the case, we can speak of something like a definite boundary for the material universe, if not perhaps for the universe as a whole.

There is one more thing I must tell you, young man, and that is
that the space between the stars appears to be empty or nearly so.
In this sense interstellar space is quite different from interplanetary
space. Again, Kapteyn has independently come to the same conclusion, so it's probably true. You know that the question of the absorption of starlight in space[8] has been much discussed by astronomers
and that it is actually of great cosmological significance, as illustrated
by Olbers' famous paradox of the darkness of the night sky. But according to my analysis the absorption is very small indeed, negligi-
ble I would say—it takes no less than 12,000 light-years of space to
change the apparent magnitude of a star by just one unit.

CCN How fascinating! You have provided us with a picture of the universe as we observe it, but what about its development in time? Has it always looked the same, or was it perhaps very different in the distant past?

HvS Really, this is a question that we know nothing about and which must be left to speculation—and speculation is not the business of astronomy. I think we can safely assume that the stellar universe as a whole is static, meaning that its size, structure, and composition have not changed over even very long periods of time. Of course, there are local changes on a stellar and nebular scale, but a truly global change is a different matter. In all likelihood the universe will remain the same in the far future.

CCN Okay, let's leave it at that, then. Before we proceed to the question of space itself, I wonder if you could tell me a little about your background and career, of how you came into astronomy and perhaps about some of the scientists you have met in your long life as an astronomer?

HvS Hmm, thank you for reminding me of my age, young man! But there is no need to go into details, I think, science is after all more interesting than scientists[9] and the history of science—don't misunderstand me, I find the study of the history of science to be valuable and enlightening, even edifying, but it's just not what really matters. In any case, I was born in Biala in 1849, a town which was then part of the Austro-Hungarian Empire, and after studies in Heidelberg I went to Leipzig to write my PhD thesis. Leipzig was a great place at the time, really the birthplace of astrophysics, and it was exciting to meet important physicists and astronomers such as Neumann, Bruhns, and Zöllner.[10]

CCN So you knew Zöllner?

HvS Oh yes, I knew him quite well, he was older than me but not by very much. I was in Leipzig when he published his book on the nature of comets, which created a big controversy and seriously damaged his reputation. You see, this book—it is hardly known today, I think—was wildly polemical and included attacks on some of Germany's most respected men of science, including the great Helmholtz.[11] Will you imagine, accusing Helmholtz of dishonesty and not having understood what science is all about! Oh, this was a great scandal and it only grew worse when poor Zöllner turned to speculation about a four-dimensional space as an entrance to the spiritual world. I remember observing the controversy from the sidelines, thinking that these philosophical ideas about space were absolute rubbish. Incidentally, in his book on the nature of comets Zöllner even thought that cosmic space might be curved[12] and therefore finite, did you know that?

CCN Yes, I do. I looked up his book some years ago. In fact, I mentioned it to Schwarzschild, who was not aware of it.

HvS No, he was too young. But as you may know, such ideas have recently returned in the so-called theory of relativity, where they are defended by Einstein and a few others. But take my word for it, they are no less wrong today than they were in Zöllner's time. But enough of that. I later returned to Leipzig as a lecturer and eventually, in 1882, I was appointed professor of astronomy here in Munich and director of the Bogenhausen Observatory. I replaced a Scotsman by the name of Johann von Lamont,[13] who was mostly interested in the Earth's magnetism and had rather neglected the astronomical part of the observatory. So I had to bring it up to date.

I have had several offers of other posts, but always decided to stay in Munich, which I have never regretted. Not only is it a lovely city with a fine observatory and an important academy, but over the years I also succeeded creating a school of astronomy with many excellent students. As I think I mentioned, Schwarzschild was one of them and easily the brightest. Do we really have to continue? No, let us turn to something more substantial.

CCN Right, but first, if you don't mind, I would like to know about your opinion with regard to the present conditions of doing astronomical research here in Germany. How have you managed to keep on in spite of the difficult situation?

HvS Ach, it has been difficult indeed and it still is difficult. During the war there were almost no students left and our scientific staff was greatly reduced. In fact, the operation of the observatory was left to just me and Grossmann.[14] And then, as you know, last year there was this communist revolt, right here in Munich, but fortunately it was crushed by the loyal forces. It was terrible, they murdered people — the communists, I mean. . . . It's over now, but there is still unrest and we have to fight with the inflation and lack of resources—but you know all that. And the situation is no less worrying from an intellectual point of view, for it is as if we are facing a hostile environment[15] where mysticism counts more than science, and astrology more than astronomy. It's the spirit of our time, I'm afraid, and I don't know what to do about it. Believe me, it was better in the days before the Weimar Republic. Much better.

CCN Perhaps so, I know that we live in a period with values and attitudes that are not always in agreement with a scientific world view.

HvS That's an understatement. Nonetheless, I consider it our duty to keep German science healthy in this difficult time, and so far we have succeeded. As you probably know, we are not welcome in the so-called International Astronomical Union,[16] but that's the least of my worries. If they don't want German astronomy, so much the worse for them. Although we lost the war, we have our dignity as a nation[17] to take care of. I'm sure you know what I mean. But I don't want to speak of politics.

CCN I understand that, it's a dreadful subject, so let's turn to cosmology. Your present idea of a limited stellar universe seems to agree with some arguments of a quite different nature that you proposed in about 1895 in connection with the universal validity of Newton's law of gravitation. Could you comment on this?

HvS Yes, you see, this is really a problem that goes all the way back to Newton himself and which I reconsidered in terms of strict mathematics. As far as I know, I was the first to do this. The question is about the stability of an infinite stellar universe[18] governed by the law of gravitation. It turns out that inconsistencies unavoidably turn up in such a universe. For example, the gravitational force exerted on a body by all the other masses does not lead to a unique result. And that's not all, for other pathological results appear as well, such as motions that start with a finite speed and accelerate to infinitely great speeds in a finite period of time. We can't have that, can we?

Let me see, in an old paper in some American journal[19] I formulated the problem as follows. Now, where do I have it? Ah, here it is: "Infinitely great accelerations must occur in the universe, and this would be true with any conceivable mode of mass-distribution whatever." And then I noticed the mentioned paradox with infinite speeds obtained in a finite time, concluding that it "contains within itself either an absurdity or a direct contravention of the theory of mechanics."

CCN I see, so your argument was of the type that philosophers call a "reductio ad absurdum" or a proof by contradiction?

HvS Precisely. And so I thought, what can be done about it? One possibility is to modify Newton's law at very large distances—after all, this marvellous law is nothing but an empirical generalization and we have no certain knowledge that it is absolutely true or holds good very far from the solar system. There is even the possibility that the law of attraction in one region of the universe differs from that in another region[20]—which is a speculation, of course, but I believe it's an admissible one.

And there is the possibility, perhaps not quite as speculative, that at very large distances a celestial body moves as if there were a repulsive force in addition to the attractive gravitational force—or as if gravitation were somehow absorbed in space. In my paper in the *Astronomische Nachrichten*[21] I suggested such a modified law by introducing an attenuating factor given by a certain constant which I labelled lambda. If this is done the mentioned inconsistencies will disappear. On the other hand, I never intended this modified law to be an actual law of nature, it was just meant to be an example. And I definitely did not introduce it to save the infinite stellar system of Newton, for that is a kind of system I don't believe in at all.

CCN So, you maintain that the material universe must be finite and that, after all, we need to use Newton's law as the only one which has solid empirical support. It is interesting that Einstein, in his recent cosmological model based on the theory of relativity, also introduces a cosmic repulsive force that he claims is necessary in order that the finite universe be gravitationally stable. Is there any connection?

HvS I think not! I have not studied Mr Einstein's work in any detail, but his entire line of reasoning—not to mention the conceptual foundation of his theory—is very contrary to mine. It's true that he has this constant or cosmological term, as I believe he calls it, but it has only a superficial similarity to mine. On the other hand, now you mention it, there might be some connection, for I wonder if he came to it independently[22] or took it from me—without citing my paper! After all, wouldn't you agree that in a formal sense it looks conspicuously like my lambda constant? Einstein surely must be a reader of *Astronomische Nachrichten*, and he even uses the same symbol[23] for the constant that

I used! But then perhaps people in Berlin have other standards for citing scientists than we have here in Munich.

CCN You don't think that Newton's law of gravitation is valid in an infinitely large universe, or perhaps that it is not even meaningful—but what about other laws of nature? The validity of the laws of thermodynamics has been discussed quite a lot in the past in connection with the so-called heat death, the inference that in the far future all activity in the universe will come to a halt—the universe will, in a manner of speaking, die.

HvS Yes, I know all about the heat death, you don't have to lecture me. I once gave an address to our academy[24] on these matters, which I have reflected upon from time to time. It's important to distinguish clearly between what can be answered scientifically and what cannot, or what is physics and what is metaphysics. We have to be very careful about that. The so-called laws of nature are not absolute or divinely given laws, they are our best attempts to summarize regularities in nature as known empirically—no more than that. Now, thermodynamics is no exception, and the laws of energy conservation and increase in entropy simply lose their meaning if they are extrapolated to an infinitely large system or to the universe as a whole.

CCN So, you're saying that the laws of thermodynamics are invalid on a cosmological scale?

HvS Yes, that's what I'm saying. They make no sense on this scale. I'm not alone in this view, which has been cogently argued by a respected physicist from St Petersburg,[25] Chwolson is his name. Moreover, the much-discussed second law is really to be understood in a probabilistic sense, which undermines the strict irreversible nature that is usually associated with it. So there is no good reason to believe that the universe spontaneously evolves from a more to a less ordered state— that cosmic entropy continually increases. As I see it, the opposite tendency is not only a possibility,[26] but it may even have more in its favour than the doctrine of an irreversible degradation of energy.

CCN It is my impression that the whole question has been discussed as much in a religious context as from the perspective of science.

HvS Indeed it has, and especially by Jesuit scholars and other Catholics, for if cosmic entropy increases toward some maximum state it cannot have increased for an infinite amount of time; the universe must have had a beginning in time, corresponding to an absolute minimum entropy. And if the universe began there presumably must have been a creator[27]—at least, that's how the argument goes. Oh yes, this used to be a big theme in German cultural circles, but I find it profoundly unsatisfactory and scarcely meaningful. There was a young man—he may have been a physics teacher, but I have forgotten his name—who recently wrote an entire dissertation on this topic[28] of cosmology, thermodynamics, and religion, a topic which really has no scientific foundation and which I prefer to ignore. It's a waste of time. I am myself an evangelical Christian, but my faith has no need of thermodynamics—now, let us talk about something else.

CCN Certainly, what about the concept of space? After all, the multitude of stars and nebulae float around in space, but what is it they are floating around in? Some physicists are willing to assign a kind of physical meaning—even structure—to space itself, if I have understood them correctly.

HvS I'm afraid that you *have* understood them correctly, but you shouldn't believe them. Space really has no properties at all,[29] it does not exist physically in the same sense that apples and stars exist; it is merely a concept introduced to order and coordinate our sensory experiences—just as time is. There has been talk about geometries other than the ordinary Euclidean one—as I mentioned, Zöllner took the idea seriously—but the geometry of space is not something that can be determined objectively by means of experiment and observation. As the great Poincaré emphasized,[30] it's a convention. There is nothing more to it than that. We choose the geometry that allows the simplest description of nature, and it so happens that this geometry is the one first described by Euclid.

CCN But not all agree. Even Schwarzschild, your former student, once considered the possibility of astronomy in non-Euclidean space, and then there are the modern ideas of relativity.

HvS Yes, yes, I know, but I'm not so sure that Schwarzschild took it seriously, it was more like a game, and he certainly wasn't committed to it. And with regard to the curved space of relativity theory, or what they call space–time, I find it inadmissible. Space is surely three-dimensional and the sum of angles in any triangle, however big, is surely 180°. Are we really going to repeat the follies of Zöllner?

CCN I assume that was a rhetorical question. At any rate, I would very much like to hear about your opinion of Einstein's theory of relativity and the whole relativity business which nowadays is such a heated and controversial topic. I understand that you have been somewhat involved in the business, even before Einstein came up with the final theory of gravitation.

HvS Hmm, let me be frank, I much dislike Einstein's theory.[31] I guess I'm a conservative, you may even call me an anti-relativist, but my attitude is solidly based on scientific reason and I am certainly not the only one to question the scientific legitimacy of the theory of relativity. My son is a physicist in Greifswald;[32] he's about your age and he confirms that many of his colleagues reject Einstein's theory as artificial and unnecessary. He doesn't believe in it himself. I'm in good company.

Alas, we live in a time of unrest and confusion, and I have to admit that there are a few esteemed physicists who are in favour of the Einstein theory. One of them is Professor Sommerfeld here in Munich,[33] with whom I have sometimes discussed the issue; and shortly before his untimely death Schwarzschild became infected by the relativistic bacterium. I can't help wondering that, had he lived today, he would have realized his mistake.

CCN I know Einstein's ideas are controversial, but perhaps it will just take some time to absorb them and accommodate to them? When Newton postulated a force of gravity that mysteriously propagates through space, he too . . .

HvS Yes, I know, but this is different. What matters is that the new theory is very strange, really bizarre, from a conceptual and methodological point of view. In fact, it is so strange that it can only be accepted if it is supported by very strong empirical evidence, and no such evidence exists. Personally I would resist the theory even if the evidence existed, . . . but that is hypothetical. I want to make it clear that I have never criticized relativity in public or attacked Einstein personally, such as some physicists in Berlin have. There is the so-called *Arbeitsgemeinschaft deutscher Naturforscher* [Syndicate of German Scientists], which recently held a big meeting in Berlin[34] to counter the unhealthy influence of Einstein and his physics, but I have nothing to do with this organization. The controversy is about science, and it should be settled by scientific means. My son agrees.

CCN In the centre of this debate is Einstein's very accurate explanation of the anomalous motion of Mercury's perihelion,[35] a problem that has haunted astronomy for a long time and which relativists claim to be a solid confirmation of Einstein's theory.

HvS Precisely, but they are wrong and so are Gehrcke and the Berlin anti-relativists. A few years ago they unearthed and reprinted an old paper by Gerber,[36] a schoolmaster I think, which contains an expression that happens to be the same as Einstein's result for the perihelion advance. As soon as I saw it, I pointed out a crude mathematical mistake that makes Gerber's result nothing but a coincidence.

Of course this shouldn't be taken as support of the relativistic explanation, because the Mercury anomaly can be explained in a much more natural way by assuming the existence of small dust particles in interplanetary space, such as I did in a paper of 1906.[37] The zodiacal light strongly suggests that such dust particles exist abundantly near the Sun, and so the Mercury problem was really solved by me a decade before Einstein's theory. I still believe that this or something like it provides the correct explanation not only of Mercury's motion but perhaps also of the bending of starlight around the Sun, should the results of the British eclipse expedition[38] be confirmed. But it's too early to tell.

CCN There is also a third test of the general theory of relativity, the prediction of a gravitational redshift.[39] My understanding is that this question is as yet undecided, but I'm not sure.

HvS It's almost certainly a chimera, a mathematical result with no foundation in reality. Actually, during the war I became involved in a small but unpleasant controversy concerning this question. To put it briefly, an astronomer by the name of Freundlich,[40] a friend of Einstein's I believe, not only claimed to have refuted my dust hypothesis but also to have demonstrated the relativistic gravitation effect in the light from certain kinds of stars. Naturally I had to respond to that paper, which I did by showing that it contained several serious errors and that the correct calculations in no way lead to a conclusion favourable to relativity. So what happened? Freundlich of course had to admit his error, but he nonetheless maintained that his study was consistent with Einstein's prediction—would you believe it? On top of that, he didn't have the decency to acknowledge me as the source of the correction—really, I had never before come across such a flagrant case of scientific misconduct. Freundlich *wanted* to prove the gravitational redshift, and when I pointed out that his reasoning was incorrect he just tailored his data and calculations in support of his preconceived conclusion! Do you see what I mean?

CCN Yes, I do.

HvS Good. Then you will also understand that I had to intervene, not for personal reasons but for the sake of the integrity of science. Let me quote from my paper[41] from, let me see, yes, from January 1916: "It can be established not only that there is no indication of the presence of a gravitation effect, but what is more, that there is only a complete contradiction of the latter." My critique was methodological and ethical, not aimed at Einstein's theory, and I said so explicitly. But, in all confidentiality, I wouldn't be surprised if Einstein was somehow involved in Freundlich's fraud and that the two had coordinated it. One never knows. In any case, I think that I showed convincingly that Mr Freundlich had acted dishonestly, something which even Sommerfeld and Schwarzschild[42] had to admit. Unfortunately my just intervention had no major effect on the relativistic fad, which Freundlich irresponsibly continues to campaign for.

There is an American astronomer, whose name I cannot remember, who has done some very careful work on the solar spectrum without finding the slightest evidence for a gravitational redshift. The question is not undecided, as you suggest, it has been proved observationally that Mr Einstein's effect does not exist.

CCN Would the name of the American be St John, at the Mount Wilson Observatory?[43]

HvS Yes, you're right, the name is St John. He proved that the Einstein effect is illusory. Oh, sometimes I forget that I have passed seventy. . . . It is only now that I notice how tired I am. It has been a long evening, can we stop now?

CCN By all means, and thank you so much for talking with me, Professor Seeliger. My journey to Munich has been of great value to me.

Notes

Hugo von Seeliger was born in Biala (now Bielsko-Biala in Poland) on 23 September 1849 and he died in Munich on 2 December 1924. He studied physics, mathematics, and astronomy at the universities of Heidelberg and

Leipzig, and in 1871 he completed his PhD thesis under the Leipzig astronomer Karl Bruhns. He subsequently gained practical experience as an observer at the Bonn Observatory and by participating in astronomical expeditions. In 1882 he was appointed full professor and director of the Munich Observatory. He remained in Munich for the rest of his life. A highly esteemed theoretical astronomer, Seeliger created an important school of astronomy with many students, the most prominent of them being Karl Schwarzschild. He was also active in the organization of German science. Thus, he served as president of the Astronomische Gesellschaft from 1896 to 1921 and presided over the Munich Academy of Sciences during the difficult period from 1918 to 1923. He was a foreign member of numerous scientific societies and academies. In 1892 he was elected an associate of the Royal Astronomical Society in London.

Seeliger's scientific work ranged over a wide spectrum, from planetary astronomy to cosmological theory. His most important contribution was in the field of stellar statistics, which in the first decade of the twentieth century led to a new picture of the distribution of stars in the Milky Way. But he also dealt with a variety of other subjects, including photometry, celestial mechanics, Saturn's ring system, cosmic dust masses, and the nature of novae. An astronomer in the classical tradition, he ignored astrophysics. Much of his work was based on advanced mathematical methods. In addition to his purely scientific work, he also dealt with astronomical and cosmological questions from a more general and philosophical perspective. Thus, he suggested in 1895 that Newton's law of gravitation did not hold on a cosmological scale. Throughout his career he remained faithful to the ideals of classical physics. When Einstein's general theory of relativity changed the scene, he was unable to accept it and instead tuned into a sharp critic of the new theory.

Biographical sources: Grossmann (1925); Kienle (1925); Eddington (1925); Schmeidler (1975). Kienle (1925) includes a complete list of Seeliger's publications and the 34 doctoral theses supervised by him.

1. *what is sometimes called statistical cosmology*: The kind of statistical astronomy and cosmology pursued by Kapteyn, Seeliger, and a few others in the early part of the twentieth century was based on data derived from stellar motions and the distribution of stars. By using sophisticated mathematical and statistical techniques, their studies much improved our knowledge of the architecture of the stellar system. Seeliger published a summary of his work in a memoir of the Bavarian Academy of Sciences (Seeliger 1920). According to the historian of astronomy Erich Paul (1993, p. 56), "Seeliger ... introduced some exceptionally powerful mathematical ideas that forever changed the landscape of this field of astronomical research."

2. *Professor Kapteyn in Groningen*: The distinguished Dutch astronomer Jacobus Cornelius Kapteyn was appointed professor of astronomy at the University of Groningen in 1878. He served as director of the university's astronomical

laboratory from 1896 until his retirement in 1921, the year before his death. Unlike Seeliger he was an internationalist with good personal connections with American, British, and other foreign astronomers.

3. *improvements made by Schwarzschild*: In a paper of 1910, Schwarzschild developed the mathematical foundation of the "fundamental equation" of stellar statistics. His picture of the universe, Sun-centred and with dimensions (diameter and thickness) of about 10 and 2 kiloparsec, did not differ materially from Seeliger's. See Paul (1993, pp. 117–120).

4. *Kapteyn's recent picture*: Investigating the stellar density distribution by means of extensive data and statistical methods, Kapteyn concluded in 1920 that the Milky Way was a huge ellipsoidal stellar system with the Sun located near its centre and with the star density diminishing with increasing distance from the centre. In 1922 he revised his model to a system covering a distance in its galactic plane of about 30,000 light-years and in the direction of the galactic poles about 5000 light-years. He estimated the average star density to be about 10^{-23} g/cm³. On Kapteyn and his universe, see Paul (1986) and Paul (1993, pp. 150–157). For an informative review of the drastic changes in astronomers' views of the Milky Way and its relation to the nebulae in the first half of the twentieth century, see Smith (2006).

5. *this great debate of the universe*: The so-called "great debate" between Harlow Shapley and Heber Curtis took place at the Smithsonian Institution in Washington, DC, on 26 April 1920. Whereas Shapley argued that the Milky Way was enormous and made up almost the entire material universe, according to Curtis it was smaller and just one spiral galaxy of many. Curtis defended the view of an "island universe," which Shapley denied. No consensus emerged as a result of the debate. For details, see Smith (1982) and Berendzen et al. (1976, pp. 35–47).

6. *something Galilean about it*: This is undoubtedly a reference to *Dialogo*, Galileo's famous book from 1632. In this classic of science the two competing world systems (Ptolemy's geocentric system and Copernicus' heliocentric system) were discussed in the form of a dialogue between three fictional characters, Simplicio, Sagredo, and Salviati.

7. *Shapley's picture of the galactic system*: Curtis' picture of the Milky Way was in rough agreement with the one favoured by Seeliger. Shapley, on the other hand, argued that the system of stars and nebulae was about 300,000 light-years in diameter and 30,000 light-years thick. Moreover, he thought that the solar system was very far from the galactic centre. Although controversial, his daring model of the Milky Way attracted wide attention and received support from leading astronomers such as Arthur Eddington and George Hale. Seeliger did not comment in public on Shapley's system, but he was convinced that it was based on distances to the stars that were much too large. See Litten (1992, p. 140).

8. *the absorption of starlight in space*: This topic is also mentioned in Chapter 1. Kapteyn derived an upper limit of 0.0016 magnitudes per parsec, but thought that its actual value was negligibly small. By 1920 most astronomers found it justifiable to disregard the existence of interstellar matter. The situation changed a decade later when Robert Trumpler, a Swiss-American astronomer at Lick Observatory, provided conclusive evidence for interstellar absorption corresponding to a change in apparent magnitude of 0.67 per kiloparsec. The story of the radical change in astronomers' attitude to interstellar obscuring matter is detailed in Seeley and Berendzen (1972).

9. *science is after all more interesting than scientists*: Seeliger's attitude was, and still is common among scientists. According to Yakov Zel'dovich and Igor Novikov, two pioneering Russian cosmologists, "the history of the Universe is infinitely more interesting than the history of the study of the Universe." (Zel'dovich and Novikov 1983, preface).

10. *Neumann, Bruhns, and Zöllner*: Carl G. Neumann, a German mathematical physicist, professor at Leipzig University since 1869; Karl C. Bruhns, a German astronomer and director of the Leipzig Observatory; Johann C. F. Zöllner, a German astrophysicist, professor of physical astronomy at Leipzig University since 1872.

11. *the great Helmholtz*: The famous German physicist and medical doctor Hermann von Helmholtz was not only recognized as an brilliant scientist, but also had enormous authority and power in imperial Germany. Indeed, Helmholtz's name became synonymous with science itself.

12. *Zöllner even thought that cosmic space might be curved*: In his book on comets from 1872, Zöllner suggested a finite universe in closed space (see also Chapter 3). A few years later he turned to spiritualism, advocating a four-dimensional space. For his cosmological model, the scandal caused by his book, and his ideas about a spiritual world, see Kragh (2012a).

13. *I replaced a Scotsman by the name Johann von Lamont*: From 1835 to 1879 the Royal Observatory of Bogenhausen was run by Lamont, who had come to Germany as a young man. Lamont soon discontinued observations with the large refractor, turning instead to studies of geomagnetism. For a brief history of the Munich Observatory, see <http://www.usm.uni-muenchen.de/Geschichte_en.php>.

14. *just me and Grossmann*: The reference is to the astronomer Ernst Grossmann, who in 1919 was promoted to a professorship at Munich. He was no relation to the mathematician Marcel Grossmann, who collaborated with Einstein in the early development of the theory of general relativity and introduced him to Riemannian geometry. For the difficult situation of the Munich Observatory during and shortly after World War I, see Litten (1992).

15. *we are facing a hostile environment*: "Wide circles of the educated or half-educated public are attracted more by astrology than by astronomy," Sommerfeld complained. He added that, "in Munich probably more people get their living from astrology than are active in astronomy." Quoted in Forman (1971, p. 13), which offers a penetrating analysis of the intellectual environment in the Weimar Republic.

16. *the so-called International Astronomical Union*: The International Astronomical Union (IAU) was established in July 1919 as part of the International Research Council, an organization from which not only the losers of the war were excluded but also neutral countries such as Denmark and the Netherlands. Whereas some of the neutral countries were admitted to the IAU in 1922, Germany and Austria were not. German astronomers had an alternative in the form of the Astronomische Gesellschaft, an institution that was founded in 1863 and also included many non-German members. Germany was only admitted to the IAU in 1951. See Blauuw (1994).

17. *our dignity as a nation*: Many German scientists in the Weimar Republic were willing to subordinate international scientific relations to the interests of the nation. While the astronomer Georg Struve deplored the negative effects of the boycott policy, he also maintained that German astronomers, for the sake of the dignity of the nation, should stay outside the new International Astronomical Union. German science, he wrote in 1926, "ought to regard as its most important obligation and task the maintenance of German dignity in the world." Quoted in Forman (1973, p. 172).

18. *the stability of an infinite stellar universe*: Newton realized that a finite collection of uniformly distributed stars would collapse gravitationally, but thought that in an infinite universe a "divine power" might secure stability. Later scientists came up with solutions without relying on God's power. One solution was Seeliger's modification of the law of gravitation, and another was to design non-uniform, so-called hierarchical models of the stellar universe. As late as 1922 the Austrian physicist Franz Selety developed on this basis a Newtonian alternative to Einstein's relativistic model. For Selety's work, see Jung (2005), and for the general problem of gravitational stability see Norton (1999) and Jaki (1979).

19. *some American journal*: The article was Seeliger (1898), published in *Popular Astronomy*. Contrary to most other astronomers of his time, Seeliger did not favour an internationalist outlook. As a German astronomer he published all his research work in German. According to Eddington (1925, p. 316), his articles were "buried away in the publications of the Munich Academy [and] they are too lengthy and formal to please an English reader." Eddington further noted that Seeliger "had not much personal acquaintance with foreign astronomers."

20. *the law of attraction in one region of the universe differs from that in another region*: According to Seeliger (1898, p. 547): "Though highly probable, it is far from self-evident, that the forces of attraction follow the same law in all places throughout the universe." This may be the first suggestion coming from a distinguished scientist that a fundamental law of nature varies in space. Seeliger did not follow up on the speculation. A few later suggestions of the same kind have been unfruitful.

21. *my paper in the Astronomische Nachrichten*: Seeliger (1895). His modified law of gravitation was that two masses m and M at a distance r gravitate by a force given by $F = GmM \exp(-\Lambda r)/r^2$ rather than Newton's $F = GmM/r^2$. The quantity Λ (which Seeliger denoted λ) is an attenuation factor.

22. *I wonder if he came to it independently*: Einstein was not a regular reader of *Astronomische Nachrichten*, a journal read by nearly all astronomers. Moreover, he was unfamiliar with Seeliger's paper when he presented his own pioneering paper on relativistic cosmology in February 1917. He referred to Seeliger's analysis of the Newtonian universe in the third edition of a popular book of 1917, but without mentioning the cosmological constant. See Einstein (1918, pp. 71–72). In a paper of 1919, dealing with an anomaly in the motion of the Moon, Einstein referred to Seeliger (1909), stating that he ought to have also referred to it in his 1917 paper on relativistic cosmology. However, he did not mention Seeliger's paper of 1895. See Einstein (1998, p. 557) and also an unpublished manuscript of 1931 discussed in O'Raifeartaigh et al. (2014).

23. *he even uses the same symbol*: Seeliger in 1895 and Einstein in 1917 both used the Greek letter λ (rather than the capital Λ, which is today more commonly used). Seeliger (1895, p. 134) commented: "It is obvious that λ can be chosen so small that within our planetary system Newton's law is valid to a sufficient approximation." Einstein (1952a, p. 185) similarly wrote: "This field equation, with λ sufficiently small, is in any case also compatible with the facts of experience derived from the solar system." As will be mentioned in Chapter 5, Einstein had already considered the lambda term and used the symbol λ in 1916, and then in a context that had nothing to do with cosmology. His use of the same symbol that Seeliger had chosen was a coincidence.

24. *I once gave an address to our academy on these matters*: Seeliger insisted that the meaning of the second law of thermodynamics was restricted to the statement that the entropy of an isolated system increases in time. "The entropy law itself is not consistent with an indefinite extension of its domain of applicability," he concluded (Seeliger 1909, p. 22).

25. *a respected physicist from St Petersburg*: Orest D. Chwolson was professor of physics at the University of St Petersburg. He had good connections to scientists in

Germany and was known for his massive textbook on physics, published in a German translation in five volumes between 1903 and 1913. For his view on laws of nature and their relations to cosmology, see Kragh (2008a, pp. 146–149).

26. *the opposite tendency is not only a possibility*: "If so," Seeliger (1909, p. 21) wrote, "there will be more and more ordered states in nature as time goes on and the entropy law will therefore become increasingly invalid."

27. *there presumably must have been a creator*: The "entropic creation argument" was first formulated in 1869 by the German physiologist and physicist Adolf Fick, a professor at the University of Würzburg. It was widely discussed by scientists, philosophers, and theologians, some of whom used it apologetically, as a proof of God. Only a few astronomers and physicists took part in the debate, which was predominantly of an ideological nature. For details, see Kragh (2008a).

28. *an entire dissertation on this topic*: Josef Schnippenkötter, a German Jesuit physics teacher, wrote his dissertation in 1919, a detailed discussion of the literature that included Seeliger's ideas (Schnippenkötter 1920).

29. *Space really has no properties at all*: Seeliger (1913, p. 200) warned against "the common and therefore very fatal misapprehension that one . . . is able to decide by measurement which geometry is the 'true' one, or even, which space is the one we live in."

30. *As the great Poincaré emphasized*: According to the eminent French mathematician Henri Poincaré, the laws of nature were conventions chosen by us for reasons of simplicity. In broad agreement with Seeliger he argued that astronomical measurements cannot possibly determine the geometry of space. See for example Poincaré (1892).

31. *I much dislike Einstein's theory*: On Seeliger as an anti-relativist, see Hentschel (1997). According to Eddington (1925, p. 318), he was "an uncompromising opponent of Einstein's theory" and "a man not easily moved in his opinions." In a letter to Schwarzschild of 8 December 1910, the theoretical physicist Arnold Sommerfeld referred to the reception of relativity theory in Munich: "Seeliger *hates* this new development with the whole candour and impulsivity that characterizes his nature." (Eckert and Märker 2000, p. 391).

32. *My son is a physicist in Greifswald*: Rudolf Seeliger was at the time associate professor at the University of Greifswald, where he worked on electrical conduction in gases. He was close to the circle of right-wing physicists opposed to the theory of relativity, including Johannes Stark in Greifswald, Philipp Lenard in Heidelberg, and Ernst Gehrcke in Berlin.

33. *Professor Sommerfeld here in Munich*: Sommerfeld, professor of theoretical physics at the University of Munich since 1906, is best known for his fundamental contributions to atomic and quantum theory, but he also worked on relativity physics and was a great admirer of Einstein's theory.

34. *a big meeting in Berlin*: On 24 August 1920 the short-lived *Arbeitsgemeinschaft deutscher Naturforscher zur Erhaltung reiner Wissenschaft* [Syndicate of German Scientists for the Preservation of Pure Science] organized a series of lectures against relativity in Berlin's Philharmonic Hall, suggesting that Einstein was a Jewish plagiarizer and the theory of relativity a fad. Einstein attended the meeting and responded to it in public. For sources, see Hentschel (1996).

35. *the anomalous motion of Mercury's perihelion*: This anomaly is also mentioned in Chapter 3. It had been known since 1859 that the perihelion of Mercury rotates around the Sun at a rate than cannot be fully accounted for by Newtonian mechanics. Einstein's theory resulted in a value for the motion that agreed almost perfectly with the observed value and thus made the anomaly disappear. A complete history of the subject is given in Roseveare (1982).

36. *an old paper by Gerber*: In 1902 the German physics teacher Paul Gerber suggested a theory of gravitation that differed from Newton's and according to which gravity propagated with the velocity of light. His theory explained the Mercury anomaly and in fact led to the very same expression as that obtained by Einstein. The paper was reprinted in 1916 on the instigation of the anti-relativist Gehrcke, and used as a weapon against relativity and Einstein's priority. See Roseveare (1982, pp. 137–144) and Hentschel (1996, p. 3).

37. *as I did in a paper of 1906*: Seeliger (1906). The zodiacal light is a faint glow seen in the night sky and is caused by dust particles in the solar system originating in part from the tails of comets. For a decade or so, Seeliger's zodiacal hypothesis of the Mercury anomaly was widely accepted. See Roseveare (1982, pp. 68–93).

38. *the results of the British eclipse expedition*: The famous British solar eclipse expedition of 1919 aimed to test Einstein's prediction that starlight would be deflected by the Sun's gravitational field. In November the leaders of the expedition, Frank Dyson and Arthur Eddington, announced to the Royal Astronomical Society that light was indeed deflected in accordance with Einstein's theory. However, the effect was not unambiguously demonstrated and conservative physicists did not hesitate to come up with alternative explanations.

39. *prediction of a gravitational redshift*: In 1911 Einstein predicted that a monochromatic light wave from a massive body should be received with a wavelength greater than the one emitted, the difference in wavelength depending on the difference in the gravitational fields (see also Chapter 3). His general theory of relativity led to the same expression for the gravitational redshift. The "Einstein effect" was very difficult to detect, and by 1920 it was still not verified. The first solid evidence came in 1924, when Eddington predicted from Einstein's theory a redshift in the light from the white dwarf Sirius B that was confirmed observationally.

40. *an astronomer by the name Freundlich*: Erwin Freundlich, who later changed his name to Erwin Finlay-Freundlich, was a German physicist and astronomer. In the 1910s he became associated with Einstein and made astronomical observations with the purpose of verifying the predictions of the general theory of relativity. In 1933 he was forced to leave Germany, and between 1951 and 1959 he was professor of astronomy at St Andrews University in Scotland. For details on his life and work, see Hentschel (1997). On the Seeliger–Freundlich controversy, see also Crelinsten (2006, pp. 85–94).

41. *Let me quote from my paper*: The quotation is from Seeliger (1916, p. 86).

42. *even Sommerfeld and Schwarzschild*: See Hentschel (1997, pp. 31–32). According to Sommerfeld, Freundlich's confirmation was "more or less a fraud." Seeliger's attack on Freundlich was actually aimed at Einstein, which the latter was aware of. "Tell your colleague Seeliger that he has a ghastly temperament," he wrote to Sommerfeld in late 1915. "I relished it recently in a response he directed against the astronomer Freundlich." (Eckert and Märker 2000, p. 504).

43. *St John, at the Mount Wilson Observatory*: Charles Edward St John joined the Mount Wilson Staff in 1908. His extended work to detect the gravitational redshift failed to reveal any effect. In a letter of 28 January 1918, Eddington wrote to his colleague Walter Adams, an expert spectroscopist at Mount Wilson: "St. John's latest paper has been giving me sleepless nights, chasing mare's nests to reconcile the relativity theory with the results, or vice versa. I cannot make any headway." Quoted in Crelinsten (2006, p. 111). For details on St John's attempts to confirm Einstein's prediction, see Hentschel (1993).

5

Albert Einstein's Finite Universe

Interview conducted in Haberlandstrasse 5, Berlin, on 12 November 1928. Language: German.

By the mid-1920s it had become clear to me, somewhat belatedly, that cosmology had entered a new phase that differed significantly from the subject as studied by classical astronomers such as Kapteyn and Seeliger. The new cosmology was essentially rooted in the general theory of relativity, on the basis of which Einstein had formulated a theory that he claimed was valid for the universe as a whole. Although the qualitative features of the new theory could be grasped intuitively, unfortunately it was forbiddingly complex from a mathematical and conceptual point of view. Hard as I tried, I was forced to accept that I only understood it on a superficial level. From time to time I have returned to it, with no more luck.

I had long wished to meet Professor Einstein (Figure 6), whose theory of relativity had so profoundly changed the study of the universe and physics generally. He was of course the most famous of all scientists, courted by journalists all over the world. For this reason it was not easy to get in touch with him. By the mid-1920s several shorter interviews with him had already appeared and at least one interview-based full biography. He spent much of his time abroad, and it was only in 1926 that I got up the nerve to request a meeting with him. After some time he agreed. It may have helped that I made a substantial donation to the Astrophysical Observatory in Potsdam, also known as the Einstein Tower, making sure that Einstein became aware of my generosity. When I finally met him in his home in Schöneberg in Berlin, he was still recovering from the heart condition that had forced him to cancel our originally planned arrangement. Apparently his illness did not trouble him any longer. I had suggested that we met at a restaurant nearby, but he preferred to stay in his home in Haberlandstrasse.

Figure 6 Einstein in conversation with Eddington in 1930. Photograph taken by Winnifred Eddington, Sir Arthur's sister. Source: Douglas (1956, Plate 11).

CCN Professor Einstein, you may well be the world's most courted scientist, so I am most grateful that you have allowed me this opportunity to talk with you. In preparation for our meeting I have read Moszkowski's book as well as some other interviews, but mine will be different.

AE I hope it will, for Moszkowski's book is completely unreliable[1] and you should pay no attention to it at all. I did engage in a series of interviews with him, but he distorted what I said and misused my name to write a rambling and even . . . well, even a scandalous work that in no way reflects my ideas and opinions. In the end it caused no harm, though. I just hope you are an honest person who will not misuse our conversation for your own purpose, as Moszkowski did.

CCN Absolutely not! As I said when I wrote, it won't even be published. I actually had the suspicion that this book of Moszkowski's might be as much fiction as fact, or perhaps even more fiction than fact. I have done some other interviews, including one with Schwarzschild, and they have been unproblematic, so I can promise you that . . .

AE Okay, I'm convinced. Let's get on with the questions.

CCN Thank you. As I told you, I'm interested in cosmology, a branch of science that you have more or less established the foundations of, or rather have provided with an entirely new meaning. You first suggested a cosmological model based on the general theory of relativity[2] in 1917, but even earlier you applied your theory to astronomical problems. I'm thinking of course of your calculation of Mercury's anomalous motion around the Sun.

AE Oh yes, that was an eye-opener because it told me that the theory was right. I remember it very well, for I was quite excited—it was as if nature had spoken to me,[3] nothing less. This is a rare experience for a scientist, and a most beautiful one.

CCN Right, and how did the idea of a cosmological model occur to you? It must have been about a year later.

AE I didn't originally have the universe in mind, not at all, except perhaps in the indirect sense that I was much inspired by Mach's ideas, which are of a cosmological nature, at least sort of. When I first had my theory connecting gravitation, space, and time, it might have seemed a natural step to extend it into a theory of the structure of the universe. Natural perhaps, but that's only if seen in retrospect. It was actually de Sitter[4] who suggested it to me, possibly in the autumn of 1916, yes, I think it was. And even earlier Schwarzschild had suggested something along the same lines, but I didn't immediately take it in.

CCN And when you did?

AE Then, when I began working on it, it turned out to be a terribly difficult business.[5] It caused me a big headache. I had to think very hard until I came up with the solution of a closed spherical universe that remains stable in spite of the gravitational attraction between all the stars. I realized that this would not be the case for a homogeneous distribution of matter throughout infinite space, and also that the spatially closed continuum was not enough—I was forced to add a term to the field equations that acts as a repulsive force counteracting the attractive force. So I resurrected a constant[6] that I had briefly considered the year before in connection with the field equations of gravitation in the absence of matter. I thought that . . .

CCN Please wait, are you saying that you had the cosmological constant even before you applied your theory to the universe?

AE Yes, sort of, but at the time it was just a mathematical quantity for which I could find no use in physics, so I dismissed it. I think I relegated it to an obscure footnote. It's no wonder that you didn't notice it.

CCN This is new to me, and very interesting. Did you use the symbol λ for this useless constant?

AE Yes, I did, but then in 1917 I discovered that it was not so useless after all. On the contrary, it was essential to my cosmological model. The lambda term, which had now become the cosmological term, wasn't really justified theoretically, but I needed it and it had the advantage that I could then write down a simple expression for the mean density of matter in the universe. This provided the model with at least some connection to astronomical data. On the other hand, . . .

CCN Not so fast, professor, please. I would like to know your present opinion about this constant and also if you still subscribe to the picture of the universe you proposed more than a decade ago?

AE That's two questions . . . or perhaps it's not. Hmm, I was quite vague about the picture back in 1917, which was probably wise, for at the time I had some wrong ideas about the observed density of matter[7] and generally little knowledge about the work of the astronomers— I think I alluded to that. For example, I was only vaguely aware of the nebular redshifts that are discussed quite a lot today. Thanks to Hubble, the American astronomer, we now know that our Milky Way is just one out of a huge number of nebulae, but that came later and it doesn't really change my cosmological model, which at any rate was of a universe with homogeneously distributed matter. What I then called stars, I would today call nebulae.

    The cosmological term is another and more problematic question. I don't really like it and would rather be without it. You see, it reduces the aesthetic quality of the field equations.[8] But then I can't see how to dispense with it. There has to be something like it to keep the universe stable, and it is a simple quantity that doesn't spoil the power of the equations in any technical or formal sense. I don't know, perhaps some future development will allow me to get rid of it.

CCN It strikes me as remarkable that you associate mathematical equations with aesthetics. To the minds of most people it's a rather odd pairing, or isn't it?

AE Not to my mind, nor to that of many current physicists and mathematicians. One cannot help being fascinated by elements such as simplicity, harmony, and symmetry in one's equations, in short, by their beauty. Or perhaps sublimity is a more appropriate term. Somehow nature is the realization of the simplest conceivable mathematical ideas.[9] And yet, one should not carry this talk of beauty too far, for it is mainly concerned with the mathematical part of a theory, not so much with its physical part. I have come across theories that are very beautiful[10] but which nonetheless are wrong from a physical point of view. Alas, some of my own theories belong to this class.

CCN This is really interesting, professor. I would like to follow up on it, if you don't mind, just a bit. Aesthetics is normally associated with paintings and other works of art, and it is not uncommon to suggest a link between your theories and the ideas of modernist artists, such as Picasso. Could one perhaps say that they are different aesthetic expressions for the same kind of revolutionary thought?

AE In no way. I'm aware that I am sometimes considered the Picasso of physics,[11] or perhaps it is Picasso who's considered the Einstein of art? Whatever the permutation it's plain nonsense. Let's leave it at that.

CCN Right. So, what about the cosmological constant?

AE Yes, there's also the possibility that the constant provides a link between gravitation and the structure of electrical particles,[12] something I considered a few years after having introduced it and returned to just last year, unfortunately without any real progress. I realized early on that in a formal sense the constant can be replaced by a negative pressure and thought that it might perhaps in this way be useful outside cosmology. But it didn't work out. The formal connection between the constant and a negative pressure[13] follows trivially from the equations, but the question is whether the connection is physically significant. I couldn't see how, and still cannot. There's even the possibility that the constant may have something to do with quantum mechanics, as Weyl suggested to me,[14] but it's just a speculation and I haven't looked further at it.

CCN Have you kept up with developments in cosmology since then? I can't say that I have myself, not closely, at least, but my impression is that since you and Professor de Sitter came up with the two models, the A-solution and the B-solution,[15] the area has been in a kind of stagnation. Nothing very important has taken place, no new ideas, and people still discuss which of the two solutions best represents the observed universe. It's just the impression of an amateur, but I wonder if you agree?

AE I have probably followed the literature on cosmology even less diligently than you. I haven't lost interest in the field, but for the last several years I have concentrated on extending the equations of general relativity from a theory of gravitation to one that also describes electromagnetism, and does it in a unified way that perhaps even describes elementary particles and not only fields. I have worked on this together with my mathematical assistant Jakob Grommer,[16] who has now left me, and presently I place some hope on a new approach, something I call *Fernparallelismus* [distant parallelism]. In addition, there are the problems of quantum mechanics that I discussed last year in Brussels with Bohr, and they continue to. . . . Well, as you will understand, cosmology is at present not what concerns me most.

CCN Yes, that's perhaps understandable. Nonetheless, the theoretical framework of cosmology is entirely based on your work.

AE Well, as I said, I keep an interest in the subject. With regard to your question, yes, I guess that your impression is not unfounded. At any rate, I don't believe in de Sitter's solution; after all, the universe does contain matter, and a lot of it, and that's not its only weakness. I still believe that the best model is my original one, you know, all the matter in the universe is embedded in a closed and therefore finite space, the size of which can be calculated. Actually Hubble did so a couple of years ago,[17] on the basis of the very small mean density of matter that he infers from observations. I have been invited to California, but have not yet had an opportunity to meet him. But recently I had the opportunity to reconsider the whole question in connection with an article I agreed to write for an English encyclopaedia.[18] What I say there doesn't differ from what I said a decade ago: the universe is finite in space and infinite in time. That's the punch line.

CCN Right, your universe is closed and finite, with no possibility of going beyond it. We are prisoners of the universe we live in.

AE Of course we are.

CCN Yes, but isn't it possible that there are other closed universes,[19] independent of ours but more or less of the same kind? Some of them might have a larger radius of curvature, others a smaller radius, and others again may have a different matter density. And why rule out the possibility of intelligent creatures in some of those hypothetical universes?

AE One can imagine it, of course, just as well as one can imagine mermaids or unicorns, but not everything imaginable belongs to the world of science. These other universes would not be part of a common space–time. If they were, they would merely be distant regions within our universe.

What you're thinking of, if I understand you correctly, is the idea of multiple universes[20] that are causally disconnected, but this is close to nonsense. They would forever be caught off from ours and we wouldn't be able to establish their existence, not even in principle. Some of these imagined universes might be hospitable to intelligent life, if not necessarily of the kind we know, but obviously this is pure speculation. You're asking me if one should accept the existence of something about which we cannot possibly have empirical evidence. Speaking as a physicist, my answer is a resounding no. Had I been a novelist my answer might have been different.

CCN But it's still possible, isn't it?

AE Please Mr Nielsen, don't test my patience.

CCN Okay. Let's forget about those other worlds, then. With regard to what I said earlier about the A- and B-solutions, there have been some recent suggestions that since both models are inadequate, each in its own way, then one should perhaps try to reconcile the advantages of the A-solution with those of the B-solution. I believe Eddington was the first to say something like that.[21]

AE The advantage of de Sitter's model, perhaps its only advantage, is that it provides a kind of explanation of the redshifts, which mine does not. However, it is far from clear that the de Sitter effect is what the astronomers have observed in their spectra. What can I say? I guess that a kind of "reconciliation," to use your phrase—or was it Eddington's?—is not out of the question. It's even desirable, but how? I met a young American mathematician[22] who studied in Göttingen and he impressed me by his thorough knowledge of relativistic cosmology. He did some work on de Sitter's theory, changing it in such a way that its line element depends on time, I think, but he was uncertain as to what it meant physically and I could offer no help. He may now have returned to America. Hmm, perhaps . . . , no, I don't think I have more to say on this subject.

CCN But isn't there the possibility that the universe as described by relativity theory is not static and that it may even be limited in time? Let me check my notes. Yes, for some years ago there was a paper in the *Zeitschrift für Physik* by a Russian mathematician called Friedmann,[23] and as I recall you even responded to it?

AE Yes, I know, my memory is poor, but not that poor. Friedmann claimed that there are more solutions to the cosmological field equations, physically meaningful solutions, that is. And he thought that they describe a universe in evolution, for example with a curvature radius growing in time or oscillating in time. I read the paper and concluded, wrongly as it turned out, that he had made a mathematical error, and I communicated my objection to the journal. It was only a year or so later that I read a letter he had sent me, explaining that his calculations were right, and then one of Friedmann's Russian friends convinced me that this was really the case. I remember it was in Leiden, after I had returned from Japan. Yes, it must have been in 1923; it was when I visited Ehrenfest.[24]

   Of course, I had to admit my error, which I did publicly in a brief note to the *Zeitschrift*. That's it. I should perhaps have given the paper more thought, but this is the kind of thing that happens. In any case, from a physical point of view his idea of evolutionary solutions was unacceptable,[25] and that's what matters. I haven't heard from him since then.

CCN And no wonder, for I was told by a Russian friend of mine that he died a couple of years ago. I think he crashed in a balloon flight![26] Anyway, so Friedmann's paper made no impact either on you or other physicists?

AE I think you're right, it was considered unimportant. But you may be interested to learn that there is a Belgian physicist—and not only a physicist, for he's also a Catholic priest, perhaps a Jesuit[27]—who entertains similar ideas. I happened to meet him last year, he approached me when I was in Brussels for the Solvay meeting,[28] and he was unaware of Friedmann's paper. It was actually during a walk in the Royal Park in the centre of Brussels, I remember. He was surprised when I told him about it, unpleasantly surprised, I guess, for he had already written up for publication his idea of an expanding universe, in fact he had already published it. Yes, of course, I now remember that I actually read it and discussed it with him.

CCN This is interesting, when I return home I will try to find out about this paper. Do you recall the name of the Belgian physicist? I ought to know it, but I don't.

AE Let me see, something with . . . could it be Lemaine? I'm so bad at names. . . . No, it's Lemaître, yes, that's the name. Lemaître. He published it in a French or Belgian journal, perhaps in the Brussels Academy, but I don't think I have it any longer. You might contact my friend de Donder[29] at the Brussels Free University. He attended the Solvay conference and undoubtedly knows him—there can't be that many Belgian priests who write competently about relativistic cosmology, wouldn't you agree?

CCN Yes I do, it should be possible to look him up. Although the cosmological field equations do not give us reason to think that the universe is in a state of evolution then there is this old argument that the entropy law forces it to be decaying—the heat death argument. Is that something you have considered?

AE I once did, a long time ago, but not after I came to a universe based on general relativity. I think the field equations make up the only reliable theoretical basis for cosmology. They are more reliable than even the laws of thermodynamics, although ultimately, of course, there cannot be a conflict between them. But you should be aware that the kind of evolution—or rather devolution—thermodynamics speaks about only concerns the free energy of the universe[30] and has nothing to do with a change in the curvature or size over time. Keep in mind that Friedmann's speculations are entirely different from Clausius'. My closed universe may well be running down without changing its size, geometry, or amount of matter.

CCN Yes, I now see that I mixed up two different meanings of "evolution." There is another classical problem of cosmology, namely, whether the universe is finite or infinite. I once discussed it with Professor Seeliger in a conversation I had with him, and your model is of course finite. But it's my impression that many astronomers are unhappy with the finite universe that relativity theory offers. For example, in an essay a few years ago Charlier in Sweden[31] proposed an infinite stellar universe, and he's not the only one favouring infinity. Do you think we will ever know the answer to the question?

AE I don't see why not. It hasn't been proved observationally that space is finite, although I believe it is, but the point is that observations *can* give an answer. Contrary to what Poincaré thought, it's an empirically meaningful question.[32] I wouldn't be surprised if astronomers come up with an answer within a decade or two from now. If they do I expect it to be affirmative.

CCN Thank you, and now to something else. I know that you are also an expert in quantum theory, which makes me ask whether there might be some connection between quantum theory and cosmology.[33] It perhaps sounds a bit crazy, but I understand that a few physicists here in Germany have toyed with the idea. For example, Nernst has calculated the amount of zero-point energy[34] in the ether and used it to justify a cosmological theory he . . .

AE I know, I know, he has told me about it and he once gave a very long and very boring communication to the German Physical Society on the subject. But there's no reason to take him seriously. Not only is the zero-point energy of empty space an unphysical quantity, an artefact of quantum mechanics, but Nernst also identifies space with the classical ether and, moreover, his ideas of the universe have nothing to do with general relativity. Forget about him. I'm sorry to say that he's no longer in the vanguard of physics. Cosmology without gravitation makes no sense, and today gravitation means the general theory of relativity. Quantum mechanics may turn out to play some role in the future, but if so we have not the slightest idea of what it will be. Again, forget about it.

CCN Too bad, I thought these ideas might perhaps contribute to a better understanding of cosmology.

AE I find the approach of Stern and Lenz[35] in Hamburg to be more fruitful. You know, they consider the closed universe from the point of view of thermodynamics, trying to describe it in terms of physical processes. Lenz wrote a paper in which he discussed the equilibrium between matter and radiation in a universe of the kind I proposed in 1917, and he actually derived an expression for the radiation temperature. He found that the temperature depends on the world radius, but since we don't really know its value there seems to be no way to test the hypothesis. Nonetheless, it is not without interest and someone ought to follow up on it. But it won't be me.

CCN I wouldn't like to terminate our conversation without asking you about how the new cosmology relates to religion[36] and metaphysics. But please let me know if you are uncomfortable with the question.

AE No, I'm used to questions of this kind, so why not? For some reason people find it terribly interesting.

CCN Thanks. It seems as if the finite space of general relativity has been embraced by at least some religious leaders and advocates of an idealist world view. They suggest, if I have understood them correctly, that it opens the way for mind to have a place in the system of nature. Does that make sense to you? Does the new picture of a closed universe have any connection to spirituality and religion?

AE You are not the first one to ask this question, and I can only repeat what I have said on earlier occasions: the theory of relativity—and that includes the cosmological model based on it—is a purely scientific matter which has nothing to do with religion. It cannot provide us with guidance on morality and human values, that's something we have to find out ourselves. There really isn't more to say.

CCN Thank you. I more or less expected that answer. Still, do you believe in God?

AE I don't like this kind of question. It presupposes that God is a well-defined concept, which it's definitely not. I cannot believe in either the Hebrew or the Christian God, and yet I consider myself religious.[37]

CCN Thank you, I'll leave it at that. As a last question, if I may. I wonder if I could ask you to look into the future and offer your opinion of what we will know about the universe say ten years from now.

AE Hmm, this is another question I'm often asked, as if I were a clairvoyant—which I'm not. The honest answer is that I really have no idea, and certainly not a qualified one. There are other people who are more qualified to prophesy about the future of cosmology, de Sitter or Eddington perhaps. Theoretical cosmology will surely remain a business of the equations of general relativity which strongly suggest a closed and static universe filled with matter. I see no reason why this model shouldn't be valid in the future.

But of course there is more to cosmology than the equations, although these are what mostly interest me. There are also the observations that astronomers make with their big telescopes— I mentioned Hubble's estimate of the size of the universe and there are also the galactic redshifts. They may bring us surprises and force us to revise our picture. This is how I felt back in 1917 with regard to the cosmological term,[38] for instance, and this is how I still feel— but I'm not the right person to guess what surprises are waiting for us. My own work aims at a unified theory based on new mathematical methods, a fundamental theory of all physics, and it is of such a general nature that it is unlikely to have direct cosmological consequences.

CCN I fully understand your reluctance to prophesy about the future of science. May I ask, before we depart, about your health? I know you have had some troubles with your heart.

AE Thanks for asking. Fortunately I'm recovering. But I have learned that illness has its advantages; one learns to think.[39] I have only just begun to think, and for the next many years I intend to think as hard as I possibly can. There is so much we don't know. . . . It's no secret that I will soon turn fifty, which according to some of my colleagues should disqualify me. There is a prejudice among physicists[40]—young physicists, of course—that at this age one cannot do creative work. I once subscribed to the prejudice myself, but not any longer—which is hardly surprising.

Well, it has been a pleasure talking with you, Mr Nielsen, I only hope that our conversation has not been a complete waste of time. You really ought to have talked with someone more knowledgeable than me. Now I'll wait for my wife, she will be here shortly, we're going to attend a dinner party this evening. I think that in the meanwhile I will practice on my violin.

CCN Thank you very much, Professor Einstein, it has been a wonderful experience.

Notes

Albert Einstein was born in Ulm, Germany, on 14 March 1879 and died in Princeton, USA, on 18 April 1955. Einstein started his education at a Munich gymnasium and in 1896 he entered the Swiss Polytechnic Institute in Zürich to study physics. He graduated in 1900. Unable to get an academic position, he was appointed technical assistant at the Patent Office in Berne, Switzerland. In 1905, his *annus mirabilis*, he published a series of seminal papers, which included the special theory of relativity, the hypothesis of light quanta, and a theory of Brownian motion. He obtained his first university position in 1909, and five years later he moved to Berlin as a research professor (see Figure 6). The years he spent in Berlin were scientifically fruitful, but soon after Hitler came to power he left Germany for the United States. From 1933 until his death in 1955 he was a fellow at the Institute for Advanced Study in Princeton. He never returned to Europe.

In the period from 1905 to about 1925 Einstein did pioneering work in quantum theory. It included a theory of the photoelectric effect, leading to a law for which he was awarded the Nobel Prize in Physics in 1922. However, his primary area of research was the theory of relativity. Having developed the general theory of relativity in late 1915, he continued to explore and extend it throughout his life. In 1917 he applied the new theory to the universe, thereby providing the foundation for relativistic cosmology. His theory was initially seen as controversial, but after some of its predictions were verified it became accepted by most physicists. He also predicted the existence of gravitational waves and the gravitational lensing effect, which later became popular areas of research. Einstein did not anticipate the expansion of the universe, but soon realized that it agreed with his theory. Indeed, in 1932, together with Willem de Sitter, he proposed an expanding model with an origin in time that played an important role in later cosmology. However, Einstein's enduring occupation during his life in the United States was the search for a field theory that unified the gravitational and electromagnetic forces. With regard to the forces based on quantum mechanics, he chose to ignore them. In part for philosophical reasons, he never came to terms with the rapidly developing quantum physics.

Biographical sources: Pais (1982); Fölsing (1997); Einstein (1979). His papers and correspondence, so far up to the early 1920s, are published in *The Collected Papers of Albert Einstein* (1987–2012, various editors). A complete list of publications is available online: <http://en.wikipedia.org/wiki/List_of_scientific_publications_by_Albert_Einstein>.

1. *Moszkowski's book is completely unreliable*: The Polish-German writer Alexander Moszkowski became friendly with Einstein and conducted a series of interviews with him, resulting in a book that was published without Einstein's

consent. In fact, on the advice of his friends Einstein tried to block the publication. The book (Moszkowski 1921) appeared the following year in English, entitled *Einstein, the Searcher* and subtitled *His Work Explained from Dialogues with Einstein*. Einstein considered it incompetent and speculative, a work he would take no responsibility for. "It is inconceivable to me," he wrote to his friend Heinrich Zangger, "that you spent even a minute on Moszkowski; I haven't done so myself." (quoted in Fölsing 1997, p. 470).

2. *a cosmological model based on the general theory of relativity*: Einstein's paper in the proceedings of the Prussian Academy of Sciences is translated into English in Einstein (1952a). For analysis and comments see, for example, Kerzberg (1992) and North (1990).

3. *it was as if nature had spoken to me*: Pais (1982, p. 253), who calls the discovery "by far the strongest emotional experience in Einstein's scientific life, perhaps in all his life." In a letter of 10 November 1915 to his friend Michele Besso, Einstein wrote that his wildest dreams had been fulfilled, and he later told the Dutch physicist Adriaan Fokker that the discovery had given him palpitations of the heart.

4. *It was actually de Sitter*: For Willem de Sitter and his early influence on Einstein's ideas about cosmology, see the interview in Chapter 6. Schwarzschild's even earlier influence is mentioned in Chapter 3.

5. *a terribly difficult business*: In a letter to Paul Ehrenfest of 4 February 1917, Einstein wrote: "I have perpetrated something again as well in gravitation theory, which exposes me a bit to the danger of being committed to a mad-house. I hope there are none over there in Leiden, so that I can visit you again safely." (Einstein 1998, document 294).

6. *I resurrected a constant*: In an extensive article on the general theory of relativity published in *Annalen der Physik* in 1916, Einstein briefly considered the field equations with an added lambda term. See the footnote on p. 144 in Einstein (1952c), an English translation of the *Annalen* paper.

7. *some wrong ideas about the observed density of matter*: Einstein (1952a, p. 188) ended his paper: "Whether, from the standpoint of present astronomical knowledge, it [my model] is tenable, will not here be discussed." According to the new theory, the density ρ was given by $\rho = \Lambda c^4/4\pi G$, where Λ is the cosmological constant, c the velocity of light, and G Newton's gravitational constant. Moreover, the cosmological constant was inversely proportional to the square of the curvature radius of the universe ($\Lambda = 1/R^2$). In letters of March 1917 he adopted the estimate $\rho \approx 10^{-22}$ g/cm^3, from which he inferred a correspondingly small value for the curvature radius R, of the order 10^7 light-years. While Einstein's density estimate was about a billion times higher than the one accepted by the late 1920s, it was only slightly higher than Kapteyn's estimate of 10^{-23} g/cm^3 (see Chapter 4).

8. *it reduces the aesthetic quality of the field equations*: As early as 1919, Einstein (1952b, p. 194) wrote that the introduction of the cosmological constant

"is gravely detrimental to the formal beauty of the theory." On this subject, see Earman (2001). On the aesthetic quality of the cosmological constant, see also the correspondence between Einstein and Lemaître mentioned in Chapter 7. The "future development" that allowed Einstein to get rid of the constant was not long in the future. It came with the recognition of the expanding universe.

9. *the simplest conceivable mathematical ideas*: This is what Einstein (1982, p. 274) said in his Herbert Spencer Lecture delivered at Oxford on 10 June 1933. "In a certain sense," he added, "I hold it true that pure thought can grasp reality, as the ancients dreamed."

10. *theories that are very beautiful*: For example, in a letter to the mathematician Hermann Weyl of 5 September 1921 he referred to a new unified field theory proposed by Eddington as "beautiful but physically meaningless." (quoted in Kragh 1990, p. 287). Experts disagree on the question of aesthetic considerations in Einstein's physics. See the discussion in Engler (2005).

11. *the Picasso of physics*: Einstein was uninterested in modern art. In a letter of 4 May 1946 to an American art expert he stressed that the aim of relativity theory was to provide an objective picture of nature. "This is quite different in the case of Picasso's painting," he said. Similarities and dissimilarities between Einstein and Picasso are explored in Miller (2001), where the letter is quoted on p. 321. It is reproduced in full in Laporte (1956).

12. *gravitation and the structure of electrical particles*: Einstein's 1919 paper that attempted to forge a link between gravitation and electrical particles is translated as Einstein (1952b). In a paper of 1921 Einstein considered some important questions that were still waiting for a solution. Among them were: "Are electrical and gravitational fields really so different in character that there is no formal unit to which they can be reduced? Do gravitational fields play a part in the constitution of matter, and is the continuum within the atomic nucleus to be regarded as appreciably non-Euclidean?" (Einstein 1921, p. 784).

13. *connection between the constant and a negative pressure*: By rearranging the terms in the cosmological field equations, the cosmological constant appears as an energy density ρ_Λ and a corresponding negative pressure density $p_\Lambda = -\rho_\Lambda c^2$. Einstein was aware of the connection, but without considering it important in a cosmological context. Lemaître was the first to highlight the vacuum interpretation of the cosmological constant, such as described in Chapter 7.

14. *such as Weyl suggested to me*: Hermann Weyl was a close friend of Einstein's and a specialist in both quantum mechanics and the general theory of relativity. In a letter of 3 February 1927, he wrote to Einstein, "all the properties I have so far attributed to matter by means of Λ are now to be taken over by quantum mechanics." Einstein replied: "I cannot be happy with the half-causal and the half-geometrical, which is burying one's head in the sand.

I still believe in a synthesis between the quantum and the wave conception." (quoted in Kerzberg 1992, p. 334).

15. *the A-solution and the B-solution*: Whereas the A-solution denotes Einstein's original model of a matter-filled universe, the B-solution refers to the model proposed by Willem de Sitter later in 1917. According to this model, which, like Einstein's, was based on the cosmological field equations and retained the cosmological constant, the universe was devoid of matter and apparently static. However, as soon as a test particle was placed in the universe, it would move away from the observer. Moreover, a light wave emitted far away would be shifted towards the red end of the spectrum, although the emitter does not move relative to the observer. Time runs more slowly the farther away the emitter is from the observer, which causes the "de Sitter effect." The effect and its relationship to the observed nebular redshifts was much discussed in the 1920s but only understood with Lemaître's theory of the expanding universe.

16. *my mathematical assistant Jakob Grommer*: Grommer was a Russian-Jewish mathematician who studied in Göttingen and for a period worked as Einstein's assistant in connection with unified field theories.

17. *Hubble did so a couple of years ago*: Hubble (1926) obtained an average mass density of the universe of 1.5×10^{-31} g/cm^3, much less than previous estimates. By inserting this value into Einstein's expression for the radius and mass of the universe, he arrived at $R = 2.7 \times 10^{10}$ parsecs $= 8.8 \times 10^{10}$ light-years and $M = 9 \times 10^{23}$ solar masses. In the near future, he foresaw, "it may become possible to observe an appreciable fraction of the Einstein universe." See also the interview with Hubble in Chapter 9.

18. *an article I agreed to write for an English encyclopaedia*: His article for the 1929 edition of *Encyclopaedia Britannica* showed no trace of an evolving universe. "Nothing certain is known of what the properties of the space-time continuum may be as a whole," Einstein wrote. "Through the general theory of relativity, however, the view that the continuum is infinite in its time-like extent but finite in the space-like extent has gained in probability." Einstein (1929), quoted in Kerzberg (1992, p. 335).

19. *there are other closed universes*: Moszkowski (1921, p. 133) quotes from his conversations with Einstein: "Thus we must reckon with the finitude of our universe, and the question of regions beyond it can be discussed no further, for it leads only to imaginary possibilities for which science has not the slightest use." It is possible that Einstein actually said this, but it is also possible that the quoted comment is due to Moszkowski.

20. *multiple universes*: Einstein anticipates the modern debate over the so-called multiverse, which is described in Kragh (2011, pp. 255–290). According to critics of the multiverse, the hypothesis of many causally separate universes

escapes testing and is therefore not scientific. Advocates of the hypothesis argue that although it cannot be refuted directly by empirical means, it is nonetheless genuinely scientific because it rests on a theoretical foundation that has testable consequences.

21. *Eddington was the first to say something like that*: "It is sometimes urged against De Sitter's world that it becomes non-statical as soon as any matter is inserted in it. But this property is perhaps rather in favour of De Sitter's theory than against it." (Eddington 1923, p. 161).

22. *a young American mathematician*: The American physicist and mathematician Howard Percy Robertson was a postdoctoral student in Göttingen and Munich from 1925 to 1927. While in Göttingen he completed an important paper on relativistic cosmology in which he derived a relation between the velocity of nebulae and their distances. For the radius of the observable world he calculated $R = 2 \times 10^{25}$ m. Although he had a velocity–distance relation and referred to Slipher's redshifts, he did not conclude that the universe is in a state of expansion.

23. *a Russian mathematician by the name Friedmann*: Although often described as either a mathematician or a meteorologist, Alexander Friedmann was professor of physics at the University of St Petersburg. In 1922 he published the first full analysis of Einstein's cosmological field equations, including expanding and other evolving solutions in which the world radius depends on time. He also introduced the idea of a universe of finite age, originating in a singularity ($R = 0$) at $t = 0$. His paper is today recognized as a milestone in the history of cosmology, but at the time it was ignored. On Friedmann's important work, see Belenkiy (2012) and Tropp et al. (1993).

24. *when I visited Ehrenfest*: Between October 1922 and February 1923 Einstein was on an extended tour to Japan. He was informed about Friedmann's objection by Yurij Krutkov, a Russian physicist who stayed in Leiden in May 1923, at the same time as Einstein visited Ehrenfest.

25. *his idea of evolutionary solutions was unacceptable*: In his note of 1923 acknowledging the mathematical correctness of Friedmann's result, Einstein admitted that they showed that "in addition to the static solutions to the field equations there are time varying solutions with a spatially symmetric structure." In his draft manuscript he added that "a physical significance can hardly be ascribed" to these solutions. Einstein decided to delete the latter remark, although it clearly reflected his opinion (see Tropp et al. 1993). The letter is reproduced in facsimile in Nussbaumer and Bieri (2009, p. 91).

26. *he crashed in a balloon flight!*: Friedmann was an experienced balloonist, who for a period held the Soviet height record of 7400 m. However, his death on 16 September 1925 was caused by typhus of the stomach and not by a ballooning accident.

27. *he is also a Catholic priest, perhaps a Jesuit*: Georges Lemaître was ordained a priest in 1923, but contrary to what is often stated he did not belong to the Jesuit order. When Einstein met him, he had recently been appointed professor of physics at the Catholic University of Louvain. He presented his (later so famous) paper on the expanding universe to the Brussels Academy of Science on 25 April 1927. See also the interview with Lemaître in Chapter 7.

28. *I was in Brussels for the Solvay meeting*: The famous Solvay congress convened in Brussels from 24 to 29 October 1927 and is best known for the epic discussions between Einstein and Bohr concerning the interpretation of quantum mechanics. Lemaître (1958a) recounted his first meeting with Einstein as follows: "While walking in the alleys of the Parc Léopold, [Einstein] spoke to me about a little noticed article that I had written the previous year on the expansion of the universe and which a friend had made him read. After some favourable technical remarks, he concluded by saying that from the physical point of view that appeared completely abominable to him. As I thought to prolong the conversation, Auguste Picard, who accompanied him, invited me to go up by taxi with Einstein, who was to visit his laboratory at the University of Brussels. In the taxi, I spoke about the speeds of the nebulae and I had the impression that Einstein was hardly aware of the astronomical facts." Although Lemaître may have written the article "the previous year," it was only published in 1927.

29. *my friend de Donder*: The Belgian theoretical physicist Théophile de Donder, professor at the Université Libre in Brussels, did work on thermodynamics and general relativity theory. He was acquainted with Einstein and also with Eddington, who in a letter to de Donder of 24 December 1924 described Lemaître as "a very brilliant student, wonderfully quick and clear-sighted, and of great mathematical ability." (quoted in Douglas 1956, p. 111).

30. *the free energy of the universe*: If E is the total energy of a system, the so-called Helmholtz free energy F is defined as $F = E - TS$, where T is the temperature and S the entropy. For constant volume and temperature, growth in entropy thus corresponds to a decrease in free energy.

31. *Charlier in Sweden*: The leading astronomer Carl Charlier, professor at the University of Lund in southern Sweden, proposed a Newtonian "hierarchical" model of the universe that avoided Olbers' paradox. In lectures given at the University of California in 1924, he reviewed his model without mentioning Einstein's alternative or other ideas based on the theory of relativity (see Charlier 1925).

32. *an empirically meaningful question*: In an address to the Prussian Academy of Sciences in 1921, Einstein said: "The question whether the universe is spatially finite or not seems to me an entirely meaningful question in the sense of practical geometry. . . . I do not even consider it impossible that the question will be answered before long by astronomy." (Einstein 1982, pp. 235–239).

33. *connection between quantum theory and cosmology*: A few physicists considered such a connection in the 1920s. Apart from Stern and Lenz, the Hungarian physicist Cornelius Lanczos suggested in 1925 a version of the Einstein universe where the world radius R was related to Planck's constant (Kragh 2007b, p. 165). Neither this nor other quantum-related theories were taken seriously.

34. *Nernst has calculated the amount of zero-point energy*: In 1911 Planck introduced the idea of a quantum zero-point energy, implying that an oscillator will have energy even at the absolute zero of temperature. It can never be brought to rest. The hypothesis remained controversial until it was justified by quantum mechanics and demonstrated experimentally. Even more controversial was Walther Nernst's assumption that empty space is filled with electromagnetic zero-point energy. In a paper of 1916 he calculated the energy density of vacuum—or ether, as he saw it—to no less than 1.5×10^{16} J/cm^3, which by Einstein's $E = mc^2$ translates into 150 g/cm^3. In works from the 1920s he developed the idea into a heterodox cosmological theory of a steady-state universe. For details see Kragh (1996, pp. 151–157) and Kragh (2012b).

    Nernst's hypothesis received scant attention and soon went into oblivion. Only much later, and especially after the discovery of dark energy, did it attract some interest. Indeed, with some goodwill Nernst's ideas can be considered an anticipation of dark energy. Although they have to some extent been vindicated, not all physicists agree that vacuum can be endowed with zero-point energy and associated fluctuations.

35. *the approach of Stern and Lenz in Hamburg*: In 1926 Einstein's former collaborator Otto Stern considered a static universe filled with matter and radiation. In the same year Stern's colleague at the University of Hamburg, Wilhelm Lenz, developed Stern's idea into a model of the Einstein universe based on thermodynamic reasoning. He concluded that at equilibrium the radiation energy must equal the matter energy and that the radiation temperature would vary inversely with the square root of the world radius. Assuming the radiation temperature of space to be 1 K, he calculated the world radius to be of the order 10^{29} m. For discussion and references, see Kragh (2012b).

36. *how the new cosmology relates to religion*: The subject was discussed in *Nature* in 1921 by the British idealist philosopher H. Wildon Carr, who argued that Einstein's finite space undermined materialism and atheism (*Nature* **108**, pp. 247–248). On the other hand, some Christian leaders considered relativity theory a threat to religion and a spiritual world view. The subject is discussed in Jammer (1999).

37. *I consider myself religious*: In a letter near the end of his life, Einstein described himself as a "deeply religious non-believer." He was not an atheist, but neither did he believe in a personal God who rewarded or punished humans. In a letter of 1927 he wrote: "My religiosity consists in a humble

admiration of the infinitely superior spirit that reveals itself in the little that we, with our weak and transitory understanding, can comprehend of reality. Morality is of the highest importance—but for us, not for God." (see Kragh 2004, pp. 77–78).

38. *this is how I felt back in 1917 with regard to the cosmological term*: "One day, our actual knowledge of the composition of the fixed star sky, the apparent motions of fixed stars, and the position of spectral lines as a function of distance, will probably have come far enough for us to decide empirically the question of whether or not λ [the cosmological constant] vanishes. Conviction is a good motive, but a bad judge." Einstein to de Sitter, 13 April 1917, in Einstein (1998, document 325, p. 433).

39. *illness has its advantage; one learns to think*: This is what Einstein reportedly said on the occasion of his fiftieth birthday. The source is *Nature* **123** (23 March 1929), p. 464.

40. *a prejudice among physicists*: Most of the physicists who created quantum mechanics were in their twenties. According to a ditty, sometimes (but probably incorrectly) attributed to theoretical physicist Paul Dirac, himself a quantum prodigy, "Age is, of course, a fever chill/That every physicist must fear/He's better dead than living still/When once he's past his thirtieth year." The ditty is a paraphrase of a passage in Goethe's *Faust*.

6

Willem de Sitter and the Expanding Universe

Interview conducted in Professor de Sitter's office in the Leiden Observatory, 20 August 1933. Language: German.

Since my interview with Einstein in 1928 the situation in Germany had worsened considerably, first with economic depression and mass unemployment, and then with an extended period of political unrest including fights in the streets between communists and increasingly aggressive national socialists. On 30 January 1933 Adolf Hitler became the new Reich Chancellor of the then Third Reich. A few months later Einstein fled the country. The dramatic circumstances did not affect me personally, and for a while I thought that the situation would soon return to some kind of normality. As we now know, it did not.

In the late summer of that fateful year, 1933, I went to Leiden from Hamburg, where I was living at the time. The purpose of my trip was to interview Professor de Sitter (Figure 7), a highly esteemed astronomer and a key figure in the new cosmology based on the theory of relativity. Another reason for my journey was to visit my friend Hertzsprung, who served as associate director of the Leiden Observatory under de Sitter. During the interview Hertzsprung (EH) was actually present, but only as an observer. We had agreed that he should not intervene in the discussion. He almost kept to our agreement.

CCN Professor de Sitter, I am grateful for your willingness to speak about your views on cosmology and your role in the recent developments. Given that so much has happened during the last few years—nearly a revolution, one might say—I suggest that we focus on the expanding universe. But let's start farther back in time, perhaps with your important work on the astronomical aspects of the general theory of relativity just after Einstein proposed it. Was this your first encounter with relativity theory?

Figure 7 De Sitter (right) with Eddington, Cambridge 1930. Photograph by
A. J. Cannon. Source: Douglas (1956, Plate 12).

WdS Oh no, I had followed the development for quite some time,
several years before Einstein suggested his final equations for the
gravitational field. For example, in 1913 I became involved in a minor
controversy with Freundlich, who used measurements of double
stars to argue that the velocity of light might depend on the velocity
of the source. I proved that he was wrong[1] and that relativity theory
has nothing to fear from astronomical measurements. I also examined the Mercury anomaly before the war, although not according
to Einstein's theory but on the basis of an older relativity theory that
Poincaré and Lorentz had proposed. Nothing important came out of
it, except that it sharpened my interest in the theory of relativity.
So, when Einstein published his new gravitation theory I was well

prepared and immediately began working on it. Being from a neutral country I had the advantage that I had easy access to the latest news from Germany, and even to Einstein personally,[2] unlike my less fortunate colleagues in France and England.

CCN Yes, your papers published during the war turned out to be hugely important. Is it correct that it was actually you who suggested to Einstein that he should attempt to apply his theory to the universe at large, turning it into a cosmological theory?

WdS Well, I don't recall precisely, but I discussed the new gravitation theory with Einstein in the autumn of 1916 and it is quite possible that during these discussions we also talked about a future relativistic cosmology, possibly for the first time. He didn't have the idea originally. At any rate, Eddington was serving at the time as secretary of the Royal Astronomical Society and he invited me to produce an account of Einstein's new theory of which he and other British astronomers knew very little. So I wrote these three reports for the *Monthly Notices*, and I think they were quite important—Eddington certainly thought they were. He was very enthusiastic.

By the way, I also convinced Einstein that the geometry of his universe governed by the general theory of relativity should be elliptic rather than spherical.[3] Not only is there the problem of antipodal points in the spherical case, but it can also be shown that the gravitational action of a material point becomes infinite, meaning that the spherical model is inconsistent. So elliptic space is really the simpler case and clearly preferable from a physical point of view, which Einstein had to admit.

CCN Is this elliptic space the same that Schwarzschild used? I remember that he talked about it and that I didn't understand it because I couldn't form a picture of it.

WdS Yes, it's the same, and no wonder you couldn't form a picture of it. It cannot be visualized in the same way as spherical space can. Now, the important thing is that in the third of my reports[4] I went beyond Einstein, showing that his matter-filled universe was not the only solution to the cosmological field equations. Although the B-solution[5] corresponded to an empty universe I could relate it to Slipher's observations of nebular redshifts, which at the time was something new and rather mysterious. But I thought of the equations in the wrong way, as representing a closed universe where the

redshifts were not caused by a recession but by a particular space–
time metric of the static type. It is strange that at the time we all
thought in terms of static models, but we did. It was a kind of paradigm, something so natural that we just took it for granted.

CCN Paradigm? I'm not familiar with that term? Is it German?

WdS I don't think so, it may be of Greek origin. People sometimes use
it to denote a generally accepted theoretical framework. I find it a
useful term.

CCN Thanks, now I know a new word. Tell me, what was your position in the cosmological debate in the 1920s? Were you aware of the
ideas of Friedmann and Lemaître?

WdS No, I only came to know about their work later on—don't ask
me why it took so long. I actually met Lemaître here in Leiden[6] in
the summer of 1928, it was during the General Assembly of the
International Astronomical Union. I was very busy organizing the
conference—it was my last year as president and I was responsible
for it. I don't think Lemaître said anything to me about the expanding universe . . . no, I would have recalled it. You must understand,
I really wasn't much concerned with cosmology in the 1920s, when
I mostly dealt with others matters of a more classical astronomical
kind, such as the astronomical constants[7] and the irregularities in the
rotation of the Earth. And then I also spent a lot of time reorganizing
the observatory, endless committee meetings, raising funds for new
instruments, and so on—fortunately our friend Hertzsprung was of
great help. No, when I returned to cosmology in early 1930 I had not
been an active contributor to the field for more than a decade.

CCN But then things changed.

WdS Oh yes, and drastically so. It was a meeting of the Royal Astronomical Society in January that made me recognize, and also made
Eddington recognize, that one should look for solutions that were
not static. It was only then, or rather shortly after the meeting, that
we discovered Lemaître's paper and understood that cosmology had
to be based on the notion of an expanding universe. It was actually
Eddington who sent me a copy of Lemaître's paper,[8] urging me to
study it carefully. As he showed so convincingly, Lemaître I mean,
the redshifts are purely an effect of the expansion of space—it was a
real eye-opener and of course even more so in the light of Hubble's

new data. The game had changed, and in a most dramatic and wonderfully exciting way.

CCN But tell me, when we say that space is expanding is it *all* of space? Does it mean, for example, that the Milky Way grows bigger because galactic space expands?

WdS That's a good question, and one that I've asked myself. I thought at first that the answer is yes, that the space between the stars is indeed expanding. However, Eddington disagreed. He was convinced that the attractive force within the Milky Way annuls the expansive force, and I was forced to accept his argument.[9] So, it is only intergalactic space that is affected by the expansion. The extragalactic nebulae are small islands in the vast sea of cosmic space. They don't feel the expansion.

CCN Atoms consist mainly of empty space, but I assume they are also unaffected by the expansion. I mean, can one be sure that the distance between the electrons and the nucleus doesn't grow all the time?

WdS There's no need to worry about that. Atoms are very solid things.

CCN Thank you for the clarification. Now, would you happen to know when Einstein came to accept the expanding universe and also the possibility of a definite origin of the expansion? He suggested in the spring of 1931 a cyclic model, much like Friedmann's, but that's all I know.

WdS I may not be the right person to ask, but my guess is that it happened in early 1931, perhaps during his stay in California. I know that he discussed cosmology with the Mount Wilson astronomers, for he told me that, and there was also an article in one of the American newspapers[10] about his view of the universe. I once saw it, but I can't remember what it said and I don't have it any longer. He also told me that he had toyed with an expanding model in which the density remains constant as a result of the creation of matter.[11] Apparently he soon discarded the idea as just a failure. That's all I know. Why don't you write Einstein a postcard, asking him the question?

CCN Thanks, perhaps I should, but he has left Germany and I don't even know if he's still in Europe. I will wait some time, I think.

WdS Of course, I forgot about that. He has stayed in Belgium and is on his way to the United States, as far as I know.

CCN There is another question related to Einstein that I'm curious about, and that's about the cosmological constant. I would like to know your opinion of this somewhat controversial constant which Einstein has now declared a mistake but which continues to play a role in models of the type favoured by Lemaître and Eddington. Do you believe in it?

WdS Hmm, yes, sort of, although I prefer to think that my view is more than just a matter of belief. The constant was no less necessary to my early model than it was to Einstein's, but contrary to my esteemed colleague I continue to find it a useful, if no longer a strictly necessary, quantity. I mean, it is perfectly possible to devise realistic cosmological models without the constant, but if we want some physical explanation for why the universe expands rather than shrinks lambda seems to be the only answer.[12] The expansion is not due to radiation, as Lemaître originally suggested and as some people still seem to believe. I'm convinced they are wrong.

You may object that lambda is merely a name for a hypothetical quantity and, moreover, that we don't know the mechanism by which it makes the universe expand, but to me this is not a serious objection. I tend to consider it a fundamental and therefore irreducible constant of nature, a cosmic analogue of Planck's constant or the velocity of light. After all, we don't need a mechanism for Planck's constant—it's just there. And likewise, the velocity of light in empty space just is what it is. But I have no strong feelings about the cosmological constant, it's all a question of how useful it is and not a question of whether it is philosophically pleasing or not.

CCN So, in 1930 you immediately began to develop models of the kind suggested by Lemaître?

WdS Yes, that's right, and I quickly derived Hubble's linear relationship[13] with a recession constant that corresponded to the one found empirically by Hubble. By then, of course, I was also aware of Friedmann's papers. About a year later, it must have been in the summer of 1931, I gave a full discussion of the expanding universe,[14] including some dynamical solutions that Lemaître had not considered. It all went very fast. I remember that in those early papers I emphasized that the expansion of the universe is due to the lambda constant and not to the radiation pressure.

CCN When I looked up your papers I was fascinated by your graphical representations showing how the world radius of various model universes depends on cosmic time. Lemaître's original model is one of them, but you also considered oscillating models and models of a finite age starting in what looks like a singularity. And then you say in your first paper, let me see . . . , yes, you say that observational data cannot determine which of the curves corresponds to the history of the real universe, but that "The selection must remain a matter of taste,[15] or of philosophical preference." This is interesting. Do you still think so today, about three years later? A matter of taste?

WdS Well, yes, but let me explain what I mean. First, there is the possibility of discriminating observationally between the various world models by extending Hubble's relation between redshift and distance to very distant parts of the universe. You see, the models yield different predictions of how the redshifts relate to apparent luminosities, which was pointed out quite recently in a dissertation by one of von Laue's students.[16] So it's possible to determine the curvature of space by means of observations, but unfortunately this is more in principle than in practice. So far the range of observations is much too limited to serve as a test of the geometry of the universe. For the moment we don't have any choice but to rely on some kind of philosophical preference.

CCN Aha, so that's where philosophy enters?

WdS In a way, but only in a way. I'm not advocating a philosophical approach to cosmology. As an astronomer interested in describing the real universe in terms of the laws of physics I insist that a model of the universe must be firmly grounded in observation. It must be quite independent of any metaphysical belief. If we accept metaphysics, we have left the realm of science.[17] On the other hand, cosmology is about the universe as a whole and therefore about something that cannot be observed but only inferred by means of drastic extrapolations. There is evidently a danger in extrapolating far beyond what can be observed, what I call our cosmic neighbourhood, which is the reason why we should distinguish carefully between the universe and models of the universe. When it comes to the models, we unfortunately cannot avoid relying on our philosophical taste, at least not in the present situation with its scarcity of cosmologically relevant observations.

CCN And that's where metaphysics enters, after all?

WdS Well, not really. I accept that cosmology demands some small measure of philosophy, but in my view this is not the same as metaphysics. For example, we are more or less forced to assume that our cosmic neighbourhood is not exceptional,[18] that on a very large scale no part of the universe is privileged over any other part. This is a kind of philosophical assumption, but it is acceptable as long as one does not forget that it is nothing but an assumption. On the other hand, I find it quite *un*acceptable to introduce the uniformity assumption as an a priori principle,[19] as something that *must* be true. When I gave the Lowell Lectures in Boston[20] nearly two years ago I had a chance to reflect on the possibility of obtaining knowledge about the universe as a whole. If you are interested you may look up the published version of the lectures, which came out as a book last year.

CCN Yes, I know, it's a wonderful book. I found the last chapter dealing with the modern picture of the universe to be particularly interesting. We may return to the general situation in cosmology, but perhaps you could first say something about the model you proposed last year together with Einstein.[21]

WdS I can do that, although it's not really a model either of us is devoted to. It was in early 1932, when Einstein and I both were at the California Institute of Technology in Pasadena, and we just decided to look for a simple relativistic model that agreed reasonably well with the known observational data, namely, the Hubble recession rate and the mean density of matter in the universe. So we took the space curvature to be zero and also the cosmological constant and the pressure term to be zero, and it then follows straightforwardly that the density is proportional to the square of the Hubble constant. It gives a value for the density that is high, but not impossibly high. That's about all there was to it. It was not an important paper, although Einstein apparently thought that it was.[22] He was pleased to have a simple expanding model with no cosmological constant. That's it.

CCN This brief paper to which you attach no particular significance, it was published in the *Proceedings of the National Academy of Sciences*. I wonder if it was reviewed before it was accepted, if someone made suggestions, and if you had a chance to revise the original version?

WdS Reviewed? Accepted? No, oh no, we were just asked to provide something. They were happy to publish a paper with Einstein's name on it, no matter what kind of paper.

CCN Okay, I was just curious. Now, perhaps one can say that what is *not* in the paper is as important as what is in it. First of all, you did not write down how the scale or distance factor of the universe depends on time and you also had nothing to say about the finite age that is an integral part of the model. How is that?

WdS Our aim was simply to find a relation between the density and the expansion rate. But you are right, the model is of a finite age and of course we realized right away that the age is given by two-thirds of the inverse Hubble constant,[23] which implies an age of the universe that is less than the known age of the Earth. We did discuss this problem and also the problem of an apparent beginning of the universe, but we both felt uncomfortable with it and decided to ignore it in our small paper. Because the model has an origin in time, it doesn't mean that the same is the case with the universe. As I said, I didn't take it very seriously. Indeed, perhaps we shouldn't have published it.

CCN You also do not comment on another feature of the model, namely that it is spatially infinite and therefore presumably contains infinitely many stars, which is an old problem in the history of cosmology and one I discussed with Professor Seeliger many years ago. Was it something that you and Einstein discussed?

WdS No, I don't think so. I'm aware of the infinity problem in classical cosmology, but it was not something we talked about. But you are right, today we cannot be sure that the universe is finite.

CCN And what about the possibility of a cyclic or oscillating universe[24]? This is something you mentioned in your 1930 papers and which Einstein examined in a paper to the Prussian Academy. Could it be that the present expansion of the universe is only temporary and that in the far future it will start contracting, turning the currently observed redshifts into blueshifts?

WdS It's an interesting speculation that has already been discussed by Friedmann on a purely theoretical basis, as you probably know. But I doubt very much if it's more than just a speculation. In one of my papers in the *Bulletin*[25] I concluded that all cyclic models are incompatible with the empirical data we have of our actual universe—there just isn't enough matter in it. One needs to distinguish between the periodically recurring universe, as discussed by Friedmann and more recently by Tolman[26], and the kind of universe that Einstein examined, which isn't truly cyclic.

Eddington has a strong dislike of a periodic universe, and I must admit that I share it, but that's a purely personal idiosyncrasy.[27] Even Einstein's one-cycle model is highly problematic since such a universe apparently starts with a zero volume and ends in a similarly unphysical state—but then, as you pointed out, half of this nasty problem also turns up in the ever-expanding model that Einstein and I considered last year. It's all very disturbing, and for the moment I cannot see any way out of it except that the singularities are due to mathematical idealizations. They most likely are, but even then there's the no less nasty problem of the two conflicting time-scales,[28] which seems to . . .

CCN Excuse me, what do you mean?

WdS Well, astronomers agree that the age of stars and galaxies[29] is hundreds or perhaps thousands of times greater than the age of the Earth, of the order of a hundred billion years or more. Our theories of stellar evolution and the condensation of nebulae require ages of this magnitude, but the cosmological time-scale is given by the inverse of Hubble's constant which is much smaller, about 1.8 billion years. You see the dilemma: how can the universe be much younger than the stars and galaxies that make up the universe? The Hubble constant may not be accurately determined, but it cannot be much different from its current value and certainly not so small that it can accommodate an age of a hundred billion years. That's impossible.

CCN Excuse me again, but according to Lemaître's original model, the one of 1927, the universe has existed in an eternity of time, so how can it be younger than the stars? Surely its age is enormously longer than the hundreds of billions of years required for stellar evolution. So what's the problem?

WdS Ah, that's a common misconception. The kind of infinity appearing in Lemaître's model, or what today is often called the Eddington–Lemaître model, is a logarithmic infinity which has no real or physical significance. Although the universe has always existed, in a certain sense it began to evolve only a few billion years ago. Yes, the past is infinite but it does not provide an infinite amount of time for stellar evolution. This has been pointed out not only by me, but also by Lemaître and Eddington. You see, in the early stages of the expansion all physical processes would occur extremely slowly, the rate slowing down asymptotically to zero as the infinite past is approached—so the problem with the small cosmic time-scale remains. Do you follow me?

CCN Hmm, I'm not sure, but what you say does indeed present a dilemma, even a paradox. Have you thought of how to solve it?

WdS Oh yes, for the last couple of years I have thought about it a lot, although without finding a definite solution. I have some ideas, though,[30] and recently I discussed them in connection with various models of the expanding universe. One may imagine that the radius of the universe contracted to a small minimum a definite time ago and then started to expand again. The important point is that the time of the minimum, perhaps five billion years ago, should *not* be identified with the beginning of the universe. There are indications that at this brief era there was a kind of cosmic catastrophe, a critical event that one may picture as all galaxies in the universe passing simultaneously through the minimum space with enormous velocities.

CCN Sorry, but are you not entering metaphysics, the very realm you want to keep out of science? The scenario you outline may appear more metaphysical than physical.

WdS Well . . . I mean . . . I admit this is a somewhat speculative theory, or rather a hypothesis, but I think it's better and no more speculative than other theories, such as Lemaître's. It solves the time-scale problem by disconnecting the observed expansion of the universe and the evolution of the star systems it contains.

CCN So, while the expansion of space is of relatively recent origin, space itself and its material constituents have been there all the time?

WdS It's something like that that I have in mind, yes. Whether the universe is eternal or not, I cannot say, but it existed long before the expansion. It's just a speculation, of course, but, as I said, the opposite point of view is even more speculative.

CCN You have said somewhere that we must accept a universe with contradictory properties,[31] which seems to be an extreme view. I just don't understand it. How can one describe the universe in scientific terms and yet allow concepts and properties that are contradictory?

WDS You are not quoting me quite correctly. What I said, or at least what I meant, was that the concept of the universe is after all a hypothesis and not something that can be likened to a finite material structure such as a star. When we develop a theory of stellar composition, of course, we are not allowed to use contradictory concepts. But the universe is different, it's not something we can observe. In a sense, we are already familiar with using contradictory concepts in the realm of quantum physics, such as when physicists describe an electron as both a particle and a wave. As I see it, something similar to Bohr's complementarity principle[32] may be applicable to the universe.

More concretely, we have to accept that the change in the cosmic distance scale, as given by the recession constant, is different from the evolutionary changes in stellar systems. Perhaps they are not contradictory in a strict sense, but they are complementary. The two processes go on simultaneously and yet they are independent. It's rather vague, I admit, but then we are at a very early stage in the development of scientific cosmology.

EH Yes, and to make cosmology a physical science we first need to understand . . .

CCN Ejnar, please, we agreed . . .

EH Oh, I'm sorry.

CCN Professor de Sitter, you mentioned briefly Lemaître's theory, which I take to be a reference to his new idea of a beginning of the universe[33] in what he calls a primeval atom. As far as I understand, his theory avoids the time-scale difficulty and perhaps in a more natural way than your suggestion. Do you find it an attractive theory?

WdS No, I do not, and neither does Eddington, nor, for that matter, other astronomers. There are even those who ridicule it[34]—which I do not. It's a remarkable hypothesis, but the idea of an original atomic nucleus that exploded in a giant radioactive process is highly speculative. Lemaître speaks confidently of the "beginning of the universe" as if it were an intelligible concept, which in all likelihood it is not. Moreover, he only succeeds in solving the time-scale problem by introducing a value of the cosmological constant that is tailored to solve it. I much prefer his older model of 1927 to the newer one.

CCN But why be so afraid of the idea of a universe with a beginning in time? Doesn't yours and Einstein's model belong to the same class as Lemaître's primeval atom, except that the beginning is not pictured as a physical entity?

WdS In a formal sense, perhaps, but as I said earlier, it is probably not a good model for the real universe—we may just as well forget about it.[35] At a big meeting in London[36] two years ago Lemaître spoke eloquently about what he called his "fireworks theory" and how it might be vindicated by cosmic rays, but I wasn't convinced. That's all I can say. His philosophical taste is different from mine.

EH And from mine too, if I may add.

CCN Let's leave it at that, then. Thank you so much, professor, for your interesting comments. And also thanks to you, Ejnar.

Notes

Willem de Sitter was born in Sneek, the Netherlands, on 6 May 1872 and died in Leiden on 20 November 1934. He studied mathematics and astronomy at the University of Groningen, and in 1901 he received his doctoral degree for work investigating the satellites of Jupiter. In 1908 he was appointed professor of astronomy at Leiden University, where he stayed until his death; since 1919 he also served as director of the observatory (see Figure 7). During his early career he mostly worked in areas related to celestial mechanics and astrometry, but in the 1910s he increasingly focused on Einstein's general theory of relativity and its astronomical consequences. Apart from reorganizing the Department of Astronomy in Leiden, de Sitter was also active in the organization of Dutch and international astronomy. In 1921 he founded the journal *Bulletin of the Astronomical Institutes of the Netherlands* and from 1925 to 1928 he served as president of the International Astronomical Union, being responsible for the 1928 General Assembly in Leiden.

De Sitter was a key figure in the early development and dissemination of the general theory of relativity, which he discussed with Einstein in 1916–17 and described in three important reports to the Royal Astronomical Society. The reports formed the basis of Eddington's *Report on the Relativity Theory of Gravitation* published in 1918. In his third report de Sitter showed that there exists a solution to the cosmological field equations other than the matter-filled model proposed by Einstein. In spite of de Sitter's "B-solution" being empty, it indicated a possible explanation for the galactic redshifts, and for this reason it attracted much attention. While it was originally seen as representing a static universe, since 1930 it was understood as a model for an exponentially expanding empty universe. While the relative merits of the models of Einstein and de Sitter were much discussed in the 1920s, de Sitter only returned seriously to cosmology after he became acquainted with Lemaître's theory of the expanding universe. In works between 1930 and 1933 he examined this and other expanding models. Although he was an enthusiastic advocate of the expanding universe, he resisted models that had a beginning in time. He strongly disliked the appearance of metaphysics in cosmology, such as he saw signs of in some of the cosmological ideas of the 1930s. More than half a century after his death, "de Sitter space" became a central concept in inflationary models of the very early universe.

Biographical sources: Spencer Jones (1935); Blaauw (1975); Kragh (2008b).

1. *I proved that he was wrong*: Einstein valued de Sitter's contribution. In a letter to Freundlich in the summer of 1913, he wrote: "If the speed of light depends even in the least on the speed of the light source, then my whole theory of relativity, including the theory of gravitation, is wrong." (Einstein 1993, document 472, p. 555).

2. *even to Einstein personally*: During the summer and autumn of 1916, Einstein and de Sitter met on several occasions in Leiden. To Einstein, the Dutch astronomer was the ideal link to British scientists, who at the time tended to distrust everything that came from Germany. In a postcard of early 1917 he praised de Sitter for his work to "throw a bridge over the abyss of misunderstanding." (quoted in Kerzberg 1992, p. 99).

3. *elliptic rather than spherical*: For the difference between these two types of closed space, see Chapter 3. De Sitter (1917) gives a careful discussion of the relationship between spherical and elliptic space, arguing in favour of the latter. In a footnote on p. 8 de Sitter noted that Einstein had accepted his argument. See also Einstein (1998, document 311).

4. *in the third of my reports*: "Einstein's solution . . . implies the existence of a 'world-matter' which fills the whole universe . . . It is, however, also possible to satisfy the equations without this hypothetical world-matter." (de Sitter 1917, p. 6). The paper is reproduced in Bernstein and Feinberg (1986, pp. 27–48).

5. *the B-solution*: In de Sitter's model the radius of curvature R was related to the cosmological constant by $\Lambda = 3/R^2$. If a particle were introduced at a distance r from the origin of a system of coordinates, it would appear as moving away from the observer with an acceleration given by $\Lambda c^2 r/3$. According to de Sitter (1917, p. 26): "The frequency of light-vibrations diminishes with increasing distance . . . [and] the lines in the spectra of very distant stars or nebulæ must therefore be systematically displaced towards the red, giving rise to a *spurious* positive radial velocity." [emphasis added]. For an accessible summary of the B-model, see Nussbaumer and Bieri (2009, pp. 76–77, 195–196).

6. *I actually met Lemaître here in Leiden*: Lemaître assisted in the arrangement of the General Assembly of the International Astronomical Union in Leiden, 5–13 July 1928, but was not a delegate. The Leiden meeting was important, not least because a number of German astronomers participated, despite Germany not being a member of the union. They were invited by de Sitter. See Blaauw (1994, pp. 94–101).

7. *the astronomical constants*: The system of astronomical constants is defined by the International Astronomical Union. It is a collection of physical constants and parameters used in astronomy, such as the speed of light, the constant of gravitation, the astronomical unit, the Earth-to-Moon mass ratio, the mass of the Sun, and the angular velocity of the Earth.

8. *Eddington who sent me a copy of Lemaître's paper*: In 1927 Lemaître sent copies of his paper to both Eddington and de Sitter, but apparently neither of them read it (Nussbaumer and Bieri 2009, p. 122). Eddington sent a copy to de Sitter on 19 March 1930, writing on its front page: "This seems a complete answer to the problem we were discussing." See Smith (1982, p. 198). The problem referred to was to find an alternative to the A- and B-solutions that included matter and explained the redshifts. See also de Sitter's letter to Lemaître of 25 March 1930, reproduced in Luminet (1997, p. 303), where de Sitter expresses his admiration for Lemaître's paper.

9. *I was forced to accept his argument*: The Canadian astronomer Allie Vibert Douglas, a student of Eddington and later his biographer, worked at McGill University, Montreal, in the 1930s. She recalled: "When de Sitter was visiting McGill University, the author asked him the outcome of the disagreement. His reply was immediate—'Eddington was right: Eddington is always right!' " (Douglas 1956, p. 158).

10. *an article in one of the American newspapers*: The *New York Times* of 12 February 1931 included a detailed report of a meeting held at the Mount Wilson Observatory the previous day. According to the newspaper, Einstein said that "The redshift of distant nebulae has smashed my old construction like a hammer blow." As a result of the blow, "The only possibility is to start with a static universe lasting a while and then becoming unstable and expansion starting, but no man would believe this." Einstein apparently recognized the age paradox: "It [the universe] would only be ten thousand million years old, which is altogether too short a time. By that theory it would have started from a small condensation of matter at that time." See also Einstein's letter of 1 March 1931 to Michele Besso, as quoted in Nussbaumer (2014).

11. *the density remains constant*: Einstein's early consideration of a steady-state model, dating from early 1931, is analysed in O'Raifeartaigh et al. (2014). See also Castelvecchi (2014). The model anticipated in some respects the later steady-state theory proposed in 1948 (see Chapter 11). However, Einstein associated the formation of matter with the cosmological constant and not with a creation tensor added to the field equations.

12. *lambda seems to be the only answer*: "What is it then that causes the expansion? Who blows up the india-rubber ball? The only possible answer is: the lambda does it. . . . The expansion depends on the lambda alone. To some it may sound unsatisfactory that we are not able to point out the mechanism

by which the lambda contrives to do it. But . . . we have not succeeded as yet in finding any connection between this uncanny lambda and other fundamental constants of nature." (de Sitter 1931b, pp. 9–10).

13. *I quickly derived Hubble's linear relationship*: De Sitter (1930), published on 24 June, where he obtained a recession constant of $H = 490$ km/s/Mpc, which was quite close to Hubble's 1929 observational value of $H = 500$ km/s/Mpc.

14. *a full discussion of the expanding universe*: De Sitter (1931a), published on 7 August, where he pointed out on p. 144 "the utter impossibility . . . of reconciling the short time scale of the expanding universe with our ideas regarding the evolution of stars and stellar systems."

15. *The selection must remain a matter of taste*: De Sitter (1930, p. 218). See also de Sitter (1933a, p. 184): "The choice is largely a matter of taste—it must be made on æsthetic grounds." De Sitter's attitude was shared by Eddington and most other astronomers. Moreover, it was not limited to the 1930s. For examples from the 1950s, see Kragh (1996, pp. 222–223).

16. *a dissertation by one of von Laue's students*: Max Kohler wrote his dissertation on cosmological theory under the supervision of the German physicist Max von Laue, a Nobel laureate of 1914. He was the first to analyse in detail the possibility of discriminating between different relativistic models of the universe, but astronomers took little notice of his work (Kohler 1933). Only in the 1950s did it become possible to use observations of high-redshift galaxies as a cosmological test.

17. *If we accept metaphysics, we have left the realm of science*: De Sitter favoured an empiricist view of science where observation has priority over theory. According to him, Einstein's general theory of relativity was uncontaminated by metaphysics and essentially based on the inductive-empirical method. Incidentally, this was not Einstein's view. On de Sitter's anti-metaphysical view of science, see Gale (2005).

18. *our cosmic neighbourhood is not exceptional*: The assumption of large-scale homogeneity and isotropy came to be known as the "cosmological principle," a name first used by the British astrophysicist Edward Arthur Milne in his *Relativity, Gravitation, and World Structure* (1935). See Kragh and Rebsdorf (2002, p. 39). The general idea can be traced back to Nicholas of Cusa (Cusanus) in 1440, and it was stated as a principle by Svante Arrhenius in 1909, as mentioned in Chapter 2. In the context of relativistic cosmology it appeared implicitly in Einstein's model of 1917 and explicitly in the cyclic-expanding model he proposed in 1931. Originally calling it "the extended principle of relativity," Milne formulated it in 1933 as a fundamental postulate or axiom that necessarily must be true.

At the 1931 meeting of the British Association for the Advancement of Science, de Sitter spoke of our neighbourhood as "the part of the universe of which we can know anything with certainty." In making a theory of the universe as a whole, he said, "we must, however, adopt some extrapolation, and we can choose it so as to suit our philosophical taste." He found the cosmological principle to be "a very natural hypothesis." (*Nature* **128** (1931): 708).

19. *the uniformity principle as an a priori principle*: De Sitter undoubtedly referred to Milne's cosmological system, the essence of which was first presented in papers of 1932 and 1933. He much disliked Milne's world model because of its foundation in a priori principles and its lack of observable consequences (Gale 2005). On Milne's deductivist cosmology see, for example, Lepeltier (2006) and Kragh (2011, pp. 101–108).

20. *the Lowell Lectures in Boston*: This series of lectures, the first of which was given in 1839, is sponsored by the Lowell Institute in Boston (which has nothing to do with the astronomer Percival Lowell or the observatory he founded). The lectures of 1931 were published as *Kosmos* (de Sitter 1932), the last chapter of which was on "Relativity and Modern Theories of the Universe."

21. *the model you proposed last year together with Einstein*: Einstein and de Sitter (1932), a paper dated 25 January and appearing in the 15 March issue of the *Proceedings of the National Academy of Sciences*. It follows from the parsimonious Einstein–de Sitter model that the mean density of matter in the universe is $\rho = 3H^2/8\pi G$, where H is Hubble's constant and G is Newton's gravitational constant. With this density, later known as the critical density (ρ_c), the gravitational attraction is precisely balanced by the expansion (if ρ is greater, the expansion will be followed by a contraction). Inserting Hubble's value $H = 500$ km/s/Mpc, the two authors obtained $\rho = 4 \times 10^{-28}$ g/cm^3, which "may perhaps be on the high side, [but] it certainly is of the correct order of magnitude." The model played an important role in later cosmology, where it was often considered to be representative of relativistic evolution models. In more recent literature the density is given by the parameter $\Omega = \rho/\rho_c$, which has a value of 1 in the Einstein–de Sitter model.

22. *Einstein apparently thought that it was*: According to Eddington, Einstein told him in 1932 that "I did not think the paper very important myself, but de Sitter was keen on it." Slightly later Eddington received a letter from de Sitter, in which he said: "You will have seen the paper by Einstein and myself. I do not consider the result of much importance, but Einstein seemed to think that it was." (Eddington 1938, p. 128). See also Plaskett (1933, p. 251).

The work of Einstein and de Sitter was instigated by a paper by the young German astronomer Otto Heckmann, who in 1931 had pointed out that evolving solutions of the cosmological field equations not only referred to space being positively curved, but also to spaces of negative and zero curvature. Heckmann (1976, p. 28) later characterized the Einstein–de Sitter paper as "not very profound."

23. *two-thirds of the inverse Hubble constant*: According to the Einstein–de Sitter model, the scale factor R varies with time as $R(t) = at^{2/3}$, where a is a constant, meaning that $R = 0$ for $t = 0$. It follows that the age of the universe is $t^* = 2T/3$, where $T = 1/H$ is known as the Hubble time. The value $H = 500$ km/s/Mpc corresponds to $T = 1.8$ billion years and thus $t^* = 1.2$ billion years. In the early 1930s the age of the Earth was estimated to be between 2 and 3 billion years. But none of this was mentioned in the paper by Einstein and de Sitter.

24. *the possibility of a cyclic or oscillating universe*: On these kinds of cosmological models, see Kragh (2009). Einstein's 1931 model was not oscillatory in the strict sense, or what de Sitter calls periodically recurring, since it did not extend the cyclic behaviour to possible previous or later cycles. For details on Einstein's "forgotten" model, see O'Raifeartaigh and McCann (2014).

25. *one of my papers in the* Bulletin: This is a reference to the journal founded by de Sitter, the *Bulletin of the Astronomical Institutes of the Netherlands*. The paper was de Sitter (1931a).

26. *more recently by Tolman*: Richard Chase Tolman, a physical chemist and cosmologist at the California Institute of Technology, was a key figure in relativistic cosmology. In 1931 he concluded that a cyclic world model such as Einstein's could expand and contract in a continual series of cycles. He argued that such periodic behaviour did not contradict the second law of thermodynamics.

27. *a purely personal idiosyncrasy*: See de Sitter (1933b, p. 630). "I am no Phoenix worshipper," Eddington wrote in 1928, before the expanding universe had become a reality. He explained: "I am an Evolutionist, not a Multiplicationist. It seems rather stupid to keep doing the same thing over and over again." (Eddington 1928, p. 86).

28. *the two conflicting time-scales*: On this problem, to which de Sitter first referred in the summer of 1930, see Kragh (1996, pp. 73–79). The time-scale problem continued to haunt cosmology for decades. Although it eased in the 1950s, when it was realized that the Hubble time was much longer than previously thought, even in the 1980s it was still difficult to reconcile the age of old stars with the age of the universe. The problem also turns up in some of the later interviews conducted by CCN, such as in Chapter 10.

29. *the age of stars and galaxies*: The so-called long time-scale was established by James Jeans and others in the late 1920s. Jeans calculated the age of galaxies to be of the order of 10^{13} years, while other astronomers preferred ages of 10^{11} to 10^{12} years. The consensus was that the ages of stars and galaxies were several orders of magnitude higher than the age of the Earth. De Sitter (1932, p. 131) summarized the time-scale problem as follows: "Now astronomically speaking this beginning of the expansion took place only yesterday, not much longer ago than the formation of the oldest rocks on the earth. According to our modern views the evolution of a star, of a double

star, or a star cluster, requires intervals of time which are enormously longer. The stars and the stellar systems must be thousands of times older than the universe!"

In 1933 the Estonian astronomer Ernst Julius Öpik argued that the age of the Milky Way was not much more than three billion years, but only in the second half of the 1930s did the long time-scale lose its authority, soon to be replaced by a much shorter one of three to five billion years. Nevertheless, the time-scale problem remained.

30. *I have some ideas, though*: At a meeting of the Royal Astronomical Society of 12 May 1933, de Sitter (1933a, p. 185) suggested the hypothesis that the universe had passed through "a crisis" some three to five billion years ago: "The stars and probably the galaxies existed before this crisis, which was very short, but very vigorous, and left its marks on the galaxies, on the stars and on the solar system." See also de Sitter (1933b).

31. *a universe with contradictory properties*: "The 'universe' is an hypothesis, like the atom, and must be allowed the freedom to have properties and to do things which would be contradictory and impossible for a finite material structure." (De Sitter 1932, p. 133), and similarly in *Nature* (**128** (1931): 708).

32. *Bohr's complementarity principle*: According to this principle, which Bohr formulated in 1927, to grasp the nature of the quantum world one needs to make use of complementary but mutually exclusive viewpoints. From a classical perspective a phenomenon cannot simultaneously be described as a wave and a particle, but both concepts are equally necessary to describe quantum phenomena. In 1960 William McCrea suggested that another principle of quantum mechanics, Heisenberg's uncertainty principle, might be adapted to the context of cosmology. See Kragh (1996, pp. 240–242). Ideas of "cosmic complementarity" have also been suggested by later cosmologists.

33. *new idea of a beginning of the universe*: On the origin, reception, and development of Lemaître's primeval atom hypothesis, the first version of a physical big bang theory of the universe, see Kragh and Lambert (2007). Lemaître published his hypothesis in 1931, but it was generally ignored by astronomers or sometimes ridiculed. According to the model, the universe at first expanded at a furious speed after which the expansion rate slowed down to almost zero. This "stagnation phase" was followed by an accelerated expansion. The length of the stagnation phase, depending on the cosmological constant, provided the model with a great deal of flexibility with regard to the age of the universe. See also the interview with Lemaître in Chapter 7.

34. *even those who ridicule it*: In a lecture of 10 March 1933, John Plaskett, an astronomer at the Dominion Observatory, Ottawa, characterized Lemaître's primeval-atom hypothesis as "an example of speculation run mad without a shred of evidence to support it." He commented that "such vagaries by a competent mathematician do not tend to increase confidence in other work along that line." (Plaskett 1933, p. 252).

35. *we may just as well forget about it*: After having published his article with Einstein, de Sitter did not refer to it. In 1933 the American relativist and cosmologist Howard Robertson wrote an influential review of relativistic world models in which he specifically excluded those which have "arisen in finite time from the singular state $R = 0$." (Robertson 1933, p. 80). Although he included the Einstein–de Sitter paper in his bibliography, he did not mention it in the review. He also did not mention Lemaître's primeval-atom hypothesis.

36. *a big meeting in London*: At the centenary meeting of the British Association for the Advancement of Science in October 1931, leading physicists and astronomers were invited to a symposium discussing the broader aspects of cosmology. The subject of the symposium was nothing less than "The Question of the Relation of the Physical Universe to Life and Mind." Among the participants were Eddington, Jeans, de Sitter, Lemaître, and Millikan. Lemaître used the opportunity to speak of his new theory of the exploding finite-age universe (see *Nature* **128** (1931): 700–722). While dismissing models "that start from a zero radius," de Sitter praised Lemaître's 1927 theory as "essentially true" and "a very real and important step towards a better understanding of Nature."

7

Georges Lemaître's Primeval Atom

Interview conducted in Arthur Haas' office at the University of Notre Dame, Indiana, 1 May 1938. Language: English.

In the summer of 1934 I decided to leave Germany and settle in the United States. It was time to move on. The situation in Germany worsening, and with Roosevelt's New Deal the economic depression in the United States seemed to be on the decline. I had several contacts in the country, some of them businessmen and others scientists, which proved helpful in the transition. To the latter category belonged Arthur Haas, an Austrian émigré physicist whom I had met several years earlier in Germany. Aware of my interests, Haas kindly informed me of a forthcoming conference at Notre Dame University, where he was professor of physics. I was particularly thrilled to learn that Lemaître would be participating, which gave me the possibility of meeting one of the legendary figures of modern cosmology. I listened to a public lecture on cosmic rays given by the famous Arthur Compton, but was not allowed to listen to the scientific presentations. With the kind assistance of Professor Haas, Lemaître agreed to talk with me about his contributions to cosmology and his views concerning cosmology and religion.

Inspired by the conference in Notre Dame I got the idea of establishing some kind of private research institution devoted to cosmology and astrophysics. I was willing to use a substantial amount of my own fortune on the project, but needed more sponsors and cooperation with university departments. My efforts bore no fruit, and by early 1940 I shelved the project. To my surprise, the reason for the failure was not only a lack of sponsors but also lack of support from the scientific community. Few of the physicists and astronomers we contacted saw the need for an institution focused on cosmological research. Apparently the field was too small and considered too exotic to arouse much interest among the potential core group of researchers. So I saved my money.

Figure 8 Georges Lemaître in 1934, on the occasion of receiving the Prix Francqui. Reproduced with the permission of Archives Georges Lemaître, Université Catholique de Louvain, Louvain-la-Neuve, Belgium.

CCN I'm so happy to meet you, professor [Figure 8]. It was only ten years ago that I became aware of your name, in a conversation I had with Einstein in Berlin, but since then I have followed your work with great interest and look forward to knowing more about it. Unfortunately my French is poor, possibly even poorer than your German—and almost certainly than your Danish! So we'll speak in English and I hope we can focus on your amazing hypothesis of an exploding universe or what you call the primeval atom hypothesis. Have you come to this symposium to present your hypothesis?

GL No, I have been here for some time as a visiting professor and will soon return to Belgium. As you know, this is a major conference[1] with some very good speakers and it's possibly the first of its kind. At least we have nothing similar to it in Europe. It's something of an experiment, an attempt to explore the interface between cosmology and the new physics of elementary particles—who knows, that may be where the future lies.

CCN Yes, maybe. Now, with regard to your idea of a primeval atom, I have heard people referring to it as the hypothesis of a "cosmic egg"[2] which burst at the moment of creation. Just out of curiosity, is that a name you coined?

GL No, definitely not. I have no idea where it came from. I sometimes speak of the "fireworks theory," but of course it's just a flawed metaphor for something that cannot be described in words. Names seem to be more important to the public than they are to scientists.

CCN Speaking of names I have also been told that you have argued for a "Phoenix universe,"[3] a reference to the idea of a universe that dies only to re-emerge from its ashes, or a universe that successively expands and contracts. Is that a kind of cosmological model that appeals to you?

GL I remember that I used the expression some years ago ... of course it has a certain poetic charm, but no, I don't believe in it from either a scientific or a philosophical point of view. Now, could we speak of something more interesting than words?

CCN Of course. First, to go back in time to 1927, when you published your now so famous paper, Friedmann's equations had been around for some five years. People have been puzzled as to how you could have missed them, so now I ask you the same question.

GL Well, I was not a regular reader of the *Zeitschrift für Physik*, and I was not the only one who missed Friedmann's work. Perhaps it's no excuse, but that's the way it was. When Einstein made me aware of it, Friedmann had passed away, but I made a reference to his paper at a conference of 1929, pointing out that it included several of my own ideas.[4] On the other hand, you should be aware that Friedmann's paper was mathematical and general—he had nothing to say about the redshifts, for example—while mine was astronomical and argued for a particular model of the universe. In spite of their similarity from a mathematical point of view, our papers were really quite different,[5] but I don't want to get involved in issues of priority, which is just an undignified waste of time.

CCN I agree, but there's a somewhat related question I need to ask, and that concerns the English translation of your paper that appeared in the *Monthly Notices of the Royal Astronomical Society* and only then made your theory generally known in its original formulation. I was able to compare it with the original French version, and it turns out not to be a full translation.[6] Some of the sections are missing, and especially the one in which you describe the linear relationship and the expansion rate. When you noticed the discrepancy I imagine you must have been dissatisfied, to say the least. After all, the English translation doesn't show that you had the velocity–distance relation a few years before Hubble. Do you know who the translator was?

GL Indeed I do, and there's no mystery about it, for I did the translation myself after the editor of *Monthly Notices* had requested my permission to reprint the paper. You are actually the first to notice the discrepancy between the two versions. The reason that I left out parts of the original paper was simply that those parts were no longer of any scientific interest,[7] since they had been superseded by Hubble's data. My discussion in 1927 of the radial velocities was necessarily provisional, and why print something that has become obsolete? But I suppose I should have mentioned explicitly that the translation was not a full reproduction of the original paper. If the much better known English version hides the fact that I obtained the velocity–distance relation before Hubble[8] this is of no great concern to me. As I said, I am not much interested in questions of priority.

CCN That is very interesting. Let us now speak of the idea of the primeval atom, or your attempt to explain the creation of the world on the basis of physics[9]...

GL Stop, please, stop. I have never attempted to explain the *creation* of the world, which is something quite beyond scientific explanation. My hypothesis is a scenario of how the world *evolved* from a certain kind of beginning, or of the origin of motion in the universe. It emphatically has nothing to say about the moment of creation itself.

CCN Right, I must be more precise in my use of words. Anyway, the hypothesis is a daring and controversial one as it operates with an absolute beginning of the world—I hope the term "beginning" is appropriate—a state before which it makes no sense to ask questions about it. Could one perhaps say that before the explosion of the primeval quantum there was nothing? That the entire universe emerged from nothingness?[10]

GL No, I wouldn't say that. What I call the primeval atom I imagine as a huge atomic nucleus, or perhaps an inconceivably heavy isotope of the neutron with about the same mass as the entire universe, which after all is very different from nothingness. On the other hand, when I use the term "atom" it is clearly in a metaphorical sense that is close to the ancient Greek sense of the word as something completely undifferentiated and devoid of physical properties; it is not in the sense of the modern picture of the atomic nucleus, which is a complicated system of interacting protons and neutrons and possibly some other particles.

To say that my primeval atom is nothingness, such as Eddington has suggested, might give the wrong impression that it is unreal or merely an abstract philosophical concept. That's not how I think of it, not at all. I am convinced that it was created by God as a real thing, although one that defies scientific description in the ordinary meaning. Time and space didn't yet exist, for they only came into being when the atom divided into smaller things in the primordial fireworks process. And without time and space, there's no physics.

CCN In your note of 1931 you emphasize that the beginning of the universe is a kind of quantum process[11] or event, which you describe as a violent radioactive decay or explosion. Why is that so important?

GL It's really of crucial importance, for it's only with the indeterminacy built into quantum mechanics that we can avoid Kant's first antimony[12] and its foundation in the law of causality. When I speak of the original explosion or the birth of the universe, people often ask me what caused the explosion. That's a natural question, for in classical physics an event needs a cause to happen. But now the problem arises, for the cause always precedes the effect, and how can there be a causal something *before* the original quantum when time didn't exist? This is a genuine dilemma, but it is only valid under Kant's assumption of mechanical determinism. It doesn't occur in quantum mechanics, where processes are not conditioned by a cause. This is what I hinted at in my letter to *Nature* when I contrasted the quantum evolution of the world with a song recorded on the disc of a gramophone.[13]

CCN Hmm, I better look up what this antimony of Kant's is— philosophy has never been my strong point and I must admit that this term "antimony" is new to me. But you call it his first antimony, so I suppose he had more of them, whatever they are?

GL Yes, he actually had four antimonies, but it's only the first of them which is relevant to cosmology.

CCN Well, that's a relief. You do widen my horizons, professor. And so did Professor de Sitter, when I interviewed him, he taught me another new word, paradigm. But tell me, isn't there a slight inconsistency in the formulation of your note when you say that "the beginning of the world happened a little before the beginning of space and time"? I mean, if it's meaningless to speak of time until after the original explosion, how could the world possibly have begun a little before?

GL I see that you have done your homework, Mr Nielsen. My formulation was perhaps ambiguous, but not really inconsistent, for I didn't speak of "before" in a temporal sense. What I wanted to say is that space and time emerged from the primeval atom in a logical or conceptual sense. Does that make sense to you?

CCN Yes, I guess so. Now I wonder how you came upon this idea[14] of a universe with a beginning rather than staying with your earlier model in which the universe has expanded for ever. There was in the old days a lot of talk about the increase of entropy in the universe, and some used it to argue for a kind of beginning, but this was in a more philosophical sense, I think. Did this kind of argument inspire you, or was it perhaps new astronomical observations that forced you to change your mind?

GL I have never really made it clear to myself how I came to it, but no, it was not caused by new observations. I was certainly aware of the old idea of the heat death and of what one might call entropic creation.[15] It played some role in my thinking, but in precisely what way I don't recall. That Bohr's ideas of space and time have only statistical validity was also in my mind, and so was the wish to provide a cosmic time-scale consistent with the age of the stars.[16]

But perhaps I was primarily motivated by the existence of radioactive elements,[17] not only in the crust of the Earth but throughout the universe. How is it, I said to myself, that Hubble's time-scale is of roughly the same order as the half-lives of uranium and thorium? Could it be that our present universe is the nearly burned-out result of a previous highly radioactive universe? I also recall that I speculated about the possible transformation of light to matter, which is something that Millikan had proposed,[18] but I cannot reconstruct my thoughts any longer. They didn't follow any logical pattern and I never wrote them down.

CCN That's a shame, for it would have been valuable to later historians of science if you had kept track of them.

GL Perhaps, but why should that be important? What matters is not how a scientist arrives at his hypothesis but of what scientific value it is—how it is justified and how it fares when confronted with observations and theoretical criticism. Well, to try to answer your question, I would say that the radioactive argument and the problem of the time-scales were probably the two most important sources for my hypothesis. The explanation of the origin of cosmic rays did not enter originally, I think, at least not significantly. And Eddington's address on entropy and the beginning of the universe[19] was not what caused me to think about the primeval atom, it merely gave me an opportunity to make it public. I'm afraid that this is all I can say.

CCN Excellent, this is most interesting. Now there is one last question I want to ask you with regard to the note in *Nature*, and that concerns the author. You apparently wrote it as a private person[20] and not in your capacity as a university professor of astrophysics. Few readers of *Nature* would recognize the author as the celebrated father of the expanding universe. Was that deliberate?

GL In fact it was. The note was not really of a scientific nature, but rather meant to open the eyes of people to the possibility of a beginning of the world, contrary to Eddington's view. At the time it was nothing but a qualitative and speculative scenario, and for this reason I found it appropriate to sign the letter as a private person. About two months later the Belgian Astronomical Society arranged a meeting on the primeval atom hypothesis, and in the autumn I developed the scenario into a proper cosmological model aimed at physicists and astronomers.

CCN Yes, and that brings me to the subject of cosmic rays and their relevance for your cosmology. In your address to the British Association meeting[21] you argued that this heavenly radiation is a fossil of the primeval super-atom, and I understand that this is a line of reasoning you have pursued since then. Cosmic rays have grown into a popular topic of scientific research, such as this symposium is proof of … would you say they provide the main reason for believing in an original cosmic explosion?

GL I consider it a fairly strong argument, although not a proof, of course, but it's not the only argument. You are right; back in 1931 I suggested that cosmic rays are descendants of the original explosion, or rather that they have their origin in the early formation of the stars. If the rays come from the primordial atom, either directly or indirectly, they must consist primarily of charged particles—protons, negative and positive electrons, alpha particles, and possibly even heavier nuclei—and this is indeed the view favoured today. Millikan thought they were high-energy light quanta, or photons as some prefer to call them,[22] but his view is no longer tenable.

CCN So Millikan is out[23]?

GL Yes, I think that it's fair to say that in this regard he is—although he may not admit it. It's my impression that when he has settled on an idea he sticks to it and is unwilling to listen to arguments that speak against it. I think that tomorrow Compton will speak about the subject of cosmic rays, so perhaps you can get permission to listen to his talk. The particles of the penetrating radiation are known to have very high energies, even though the expansion of space reduces their energy, so it must have been even higher in the past and incredibly high several billion years ago. I have actually calculated the energies and trajectories of charged particles in the Earth's magnetic field. These calculations are terribly complicated, but fortunately I collaborate with Vallarta,[24] who's here today and will speak about our work. You might wish to speak with him, for he's an expert on the subject.

In any case, we feel pretty confident that the energies of cosmic rays are at least consistent with an explanation of them in terms of an explosive origin. I'm not saying that the theory is right, but it does have some support. In my view, there is no real difference between cosmic rays and ordinary matter. I would even go as far as to characterize matter as condensed cosmic rays.[25] But don't take my word for it, for Vallarta and I are pretty much alone in this view.

CCN Fascinating—matter as cosmic rays in disguise! Recently the cosmic rays have revealed a new elementary particle, something described as a kind of heavy electron.[26] How does this particle fit into your picture?

GL It's consistent with it, but on the other hand Anderson's particle doesn't provide additional support for my hypothesis and at the moment it's still rather mysterious. It's just there, for no good reason,[27] it seems. It would be more interesting if people found nuclei with a higher atomic number in the rays, say carbon or oxygen nuclei.

CCN I wonder if your ideas about the origin of the cosmic rays are somehow related to the suggestion that Baade and Zwicky came up with[28] a few years ago. As far as I recall, they also think that the rays come from enormous explosions and that they are made up of protons and even heavier nuclei.

GL I'm aware of their hypothesis, but it's quite different from mine in so far that they have in mind explosions of old stars—or what they call supernovae. Mine is a cosmological hypothesis involving energies much higher than in a supernova explosion and dating from a time billions of years ago. According to Baade and Zwicky, cosmic rays emerge from supernova processes that occur today and will continue to occur. Their universe doesn't have a beginning and on the whole I find their idea to be rather speculative. I have no confidence in it.

CCN We have already talked for quite some time, but I hope we can continue for just a few more minutes. I have on my list three more topics that I would like to hear your opinion of, I think we can do it rather quickly. They are, let me see …, yes, the size of the universe, the cosmological constant, and the relation between cosmology and religion. How does that sound? If you agree, I suggest that we start with the constant, which appears to have gone out of fashion.

GL You are right, it has gone out of fashion, but then I don't follow fashions and I see no good reason to skip it. I discussed the question with Einstein when I met him here in America, it was in Pasadena, but he wouldn't listen. . . . His objections to lambda seem to be largely of an aesthetic nature, that it makes the field equations more complicated and things like that. I won't deny that there is an aesthetic element to the issue[29]—but then my sense of scientific aesthetics differs from Einstein's.

CCN But it's surely not a matter of aesthetics alone?

GL No, it's not and it shouldn't be. Look, there are strong scientific reasons, empirical as well as theoretical, to maintain a positive lambda. First of all it provides a natural solution to the time-scale problem, as I think I mentioned, and in my model it is also essential to the stagnation phase during which galaxies were formed. Without lambda it seems very difficult to explain the formation of galaxies and galactic clusters. And that's not all, for generally it provides the relativistic equations with a greater empirical content—it makes them more flexible and convenient, which to my mind is an advantage. But Einstein's mind is different, he doesn't like this kind of flexibility.

CCN What about the physical meaning of the cosmological constant? When I interviewed Professor de Sitter some years ago in Leiden he told me that it produces a kind of negative pressure which blows up the universe and in this way explains the expansion. That's what he said, but I may not have understood him properly. Did I get it right?

GL I think so. It was a great loss that de Sitter passed away. . . . What he told you is basically correct, although there are also expanding models with a zero constant, so lambda is not indispensable, strictly speaking. When I was a guest professor in Washington about five years ago I actually looked into this question, pointing out that the cosmological constant can be understood as if empty space has a definite energy density[30] and a corresponding negative pressure. I gave a relation between lambda and the vacuum energy density, but nobody seems to have noticed it, or perhaps people were already aware of it. It's a rather obvious deduction from the field equations and I don't myself consider the paper important. On the other hand, the interpretation of the cosmological constant in terms of vacuum energy suggests one more reason why lambda should be taken seriously.

CCN Thank you, I wasn't aware of your paper. Now, to proceed to the next question, it is generally believed that the universe as described by relativity theory is of finite size, and I understand that in this question you subscribe to the orthodox view?

GL You may say so. I do believe that the universe is finite,[31] and in this case "believe" is the proper word to use, for it's a question that cannot be settled observationally, at least not at present. I'm an epistemic optimist, and in my view the universe in its totality must be comprehensible to the human mind, and it cannot be if it is infinite and filled with an infinite number of stars. I once discussed the question with Tolman,[32] who wasn't convinced, I think, but I'm far from alone. Many other astronomers share my feelings, not to mention the philosophers. Haas, our host here at Notre Dame,[33] is as convinced as I am that the universe must be of finite extension. It's not merely a matter of belief, but also of logic.

CCN Your epistemic objection to the infinite universe is interesting, for this is almost precisely what Schwarzschild said a long time ago,[34] when I interviewed him during the war. He too preferred a closed universe that can in principle be known to man.

GL Oh, really? I didn't know that Schwarzschild worked in cosmology.

CCN It was in the context of pre-relativistic, curved-space cosmology, but the problem is basically the same. I would imagine that your resistance to the infinite universe is not only rooted in epistemology but also in theology, is that right?

GL Yes, you're right, but the two aspects cannot be easily separated. The Church has traditionally ..., no, that's irrelevant. Now we may just as well go on to the last point of our conversation, the relationship between cosmology and theology, I understand, but let's keep it brief, can we? This is something I am often asked about, and I am a little tired of repeating myself and explaining things to people who are ignorant about religion. You see, many people are convinced that science and religion are inherently in conflict, and they just can't understand that a priest can do scientific studies without losing his faith. The atheists use science, or they misuse science, as a weapon against religion, and Christians of a fundamentalist inclination are no better. They tend to believe that the Bible is a textbook where all kinds of scientific questions can be looked up. They are both wrong, very wrong ...

CCN Please, go on.

GL I could, but I won't. To cut a long story short, science is about nature and the Bible is about salvation,[35] so there is neither a conflict nor a complete unity between the two. I once thought that one could understand Genesis better[36] by interpreting it from the point of view of modern physics, but it was a juvenile mistake. I soon became wiser.

CCN But you have proposed a theory of cosmology in which the universe came into being at a certain time in the past, isn't this in agreement with Genesis? Or let me frame my question a little differently: did you *want* a universe that agreed with your faith[37]?

GL No, that's a misunderstanding. My reasons for the primeval atom were scientific and not religious. They still are. That's it.

CCN I see. On the other hand, you don't explain how the primeval atom was formed or what caused its disintegration, but are you not implicitly saying that God was responsible? At any rate, some people see your theory as a kind of scientific legitimization of your faith.

GL I know, but I don't need science to justify my faith. It is a complete mistake that only reveals a lack of understanding of what Christian faith is all about. My theory doesn't support any religious view,[38] Christian or not, and it is not at all concerned with creation. I want to make this quite clear. Assume that astronomers one day find convincing evidence that there was no beginning and that my theory is therefore wrong. What then? Would it disprove Genesis? Not at all, for it wouldn't affect the message of the Bible in the slightest. The universe would still have been created by God.[39]

CCN Wait, I don't understand that. I mean, how can a universe which has existed in an eternity of time have been created? It makes no sense to me.

GL That's because you don't understand the true meaning of "creation," but it's a difficult question and I don't have either the time or the patience to teach you the necessary philosophy and theology. Yes, I maintain that there was an origin of the universe in the far past and that we can know something about this past, for example by studying the cosmic rays. But we can know nothing about how God created the universe. That's impossible in a much stronger sense than it's impossible to accelerate bodies to velocities beyond the speed of light.

In fact, according to scripture the Creator is essentially hidden,[40] and He is no less hidden in my theory of the primeval atom. To repeat, God is *not* to be found in the beginning of the universe. My cosmological theory is just outside religion and metaphysics, and it gives no reason to either believe or disbelieve in a divine Creator. I have never wanted it to be judged on any criteria other than those of science, and if some people nevertheless do it I will ignore them as the fools they are.

CCN Hmm, have you written about this? Can I look it up somewhere and try to understand it, even with no knowledge of theology?

GL Some years ago an American writer interviewed me[41] and published an article on my views about science and religion. It's a reasonable summary and you may wish to read it. I don't have the reference right now, but I think I can find it for you tomorrow. By the way, you may also try learning some theology, it doesn't harm.

CCN Well, no, I guess it doesn't. Thank you, Professor Lemaître, for your stimulating account. I will attend Compton's lecture tomorrow to learn more about the secrets of cosmic rays.

Notes

Georges Henri-Joseph-Edouard Lemaître was born in Charleroi, Belgium, on 17 July 1894 and died in Charleroi on 20 June 1966. After receiving his basic education at a Jesuit school, he enrolled as an engineering student at the Catholic University of Louvain, without finishing his studies. With the outbreak of World War I he enrolled in the Belgian army, and only in 1919 did he resume his studies, now in mathematics and physics. At the same time he enrolled as a student of theology. He was ordained as a priest in 1923 and subsequently followed a career within the Catholic church parallel with his scientific career. In 1936 he was elected to the Pontifical Academy of Sciences, where he served as president from 1960 to 1966.

Lemaître spent the year 1923–4 at Cambridge University as a student under Eddington. He then went to the United States, where he worked under Harlow Shapley at the Harvard College Observatory and prepared for a PhD at the Massachusetts Institute of Technology. Upon his return to Belgium he was appointed professor of physics in 1927, the year in which he published a seminal theory of the expanding universe explaining the galactic redshifts. It took three years for the theory to be noticed by the astronomical community. With the recognition of the expansion of the universe, Lemaître became internationally known, celebrated as a pioneer cosmologist. As a result, he was invited to the United States as a lecturer and guest professor. In 1934 he was awarded the prestigious Prix Francqui, Belgium's highest scientific distinction. He was at the height of his scientific career in the 1930s when he made important contributions to the general theory of relativity and developed his cosmological theory into a theory of the birth of the universe in a state he called the "primeval atom," the first version of a big bang theory. He suggested that cosmic rays were remnants of the original explosion that had taken place billions of years ago, but later developments in cosmic ray physics failed to support that theory.

After World War II Lemaître continued his work in cosmology, but he turned increasingly to other activities, for example celestial mechanics and applied mathematics by means of electronic computers. When George Gamow and others developed a quantitative theory of the early universe on the basis of nuclear physics, he ignored it. Although recognized as an authority in relativistic cosmology, he stayed on the sidelines in the developments that in 1965 led to the modern big bang theory.

Biographical sources: Lambert (2000); Farrell (2005); Holder and Mitton (2013). Stoffel (1996) includes a complete bibliography of Lemaître's publications.

1. *a major conference*: The symposium on "The Physics of the Universe and the Nature of Primordial Particles" was held at the University of Notre Dame, Indiana, on 2–3 May 1938. Arranged by Arthur Haas, it included among its speakers William Harkins, Arthur Compton, Carl Anderson, Georges Lemaître, Harlow Shapley, and Manuel Vallarta. See the report in *Science* **87** (1938): 487–490.

2. *the hypothesis of a "cosmic egg"*: In the literature from the 1950s to the present it is sometimes stated that Lemaître referred to his idea of a primeval atom as the "cosmic egg" (Kragh 2013d). There is no basis for that claim, but apparently it was already known in the late 1930s.

3. *you have argued for a "Phoenix universe"*: In a paper of 1933 and on a few later occasions Lemaître briefly discussed eternally oscillating cosmological models, associating them with the legend of the Phoenix bird (Kragh 2004, p. 140). Although admitting that they have "an indisputable charm," he did not advocate such models. Being a Catholic priest, it would have been most surprising if he had adopted such a picture of the universe, traditionally associated with materialism and atheism.

4. *it included several of my own ideas*: In a footnote to Lemaître (1929, p. 216), he referred to Friedmann's theory and pointed out that included "most of the notions and results" of his own work of 1927.

5. *our papers were really quite different*: Although Lemaître did not care much for priority, he was eager to distance his own work from that of Friedmann. In a book review of 1950 he pointed out what were his own contributions: "If my mathematical bibliography [of 1927] was seriously incomplete because I did not know the works of Friedmann, it is completely up to date from an astronomical point of view; I calculate the expansion coefficient. . . . Naturally, before the discovery and study of galactic clusters, there could be no question of establishing Hubble's law, but only to determine the coefficient. The title of my note left no one in doubt of my intentions." (quoted in Kragh and Smith 2003, p. 147).

6. *it turns out not to be a full translation*: For the English translation, see Lemaître (1931a). Until a few years ago the missing parts of Lemaître's paper attracted little attention or, if they were noticed, gave rise to claims of conspiracy and scientific misconduct. The "mystery" was only solved in 2011, when Mario Livio documented that Lemaître was the translator. See Livio (2011).

7. *no longer of any scientific interest*: "I did not find advisable to reprint the provisional discussion of radial velocities which is clearly of no actual [current] interest," Lemaître wrote to the editor of *Monthly Notices*. He added, "No formula is changed, and even the final suggestion which is not confirmed

by recent work of mine has not been modified." (quoted in Livio 2011). The final suggestion mentioned by Lemaître was that the expansion of the closed universe was caused by the accumulation of radiation in the static Einstein universe.

8. *the velocity–distance relation before Hubble*: In his paper of 1927, Lemaître found theoretically an approximately linear relationship between the recession velocity and the distance of the nebulae. For the factor of proportionality—later known as the Hubble constant—he obtained a value of 625 km/s/Mpc.

9. *the creation of the world on the basis of physics*: As a priest trained in the neo-Thomist tradition of philosophy, Lemaître was keenly aware of the difference between "beginning" and "creation." He never spoke himself of the explosion of the primeval atom in terms of the latter concept, which he used only in a philosophical and theological context. Creation or absolute beginning was for him outside the realm of science and entirely different from what he called "natural beginning." He later said about the explosion of the primeval atom: "We may speak of this event as of a beginning. I do not say a creation. Physically it is a beginning in the sense that if something has happened before, it has no observable influence on our universe. . . . The question if it was really a beginning or rather a creation, something starting from nothing, is a philosophical question which cannot be settled by physical or astronomical considerations." (quoted in Kragh 2004, pp. 147–148).

10. *the entire universe emerged from nothingness*: In his popular book *The Expanding Universe*, Eddington (1933, p. 57) commented critically on Lemaître's fireworks theory, suggesting that the "undifferentiated sameness" of the primeval atom was just another name for nothingness. On the other hand, Lemaître associated nothingness with the theological concept of *creatio ex nihilo*, creation out of nothing. On his comparison of the primeval atom with the Greek notion of "atom," see Lemaître (1958b, p. 8).

11. *the beginning of the universe is a kind of quantum process*: The title of Lemaître's letter to *Nature* was "The Beginning of the World from the Point of View of Quantum Theory." In agreement with recent ideas of Bohr and others, he stated that, "in atomic processes, the notions of space and time are no more than statistical notions: they fade out when applied to individual phenomena involving but a small number of quanta." Therefore, he argued, "If the world has begun with a simple quantum, the notions of space and time would altogether fail to have a sensible meaning at the beginning." (Lemaître 1931b).

12. *Kant's first antimony*: In his classic philosophical text *Critique of Pure Reason* (1781) Kant concluded that the concept of the universe or world is contradictory

and therefore does not cover a physical reality. In his "first antimony" he proved by means of a logical argument that the world has a beginning in time, after which he proved that it does not have a beginning in time. He repeated the argument with regard to the finite size of the universe. Kant also formulated an antimony of atomism, one of freedom and one of God.

13. *a song recorded on the disc of a gramophone*: "The whole story of the world need not to have been written down in the first quantum like the song on the disc of a phonograph," Lemaître (1931b) wrote. "The whole matter of the world must have been present at the beginning, but the story it has to tell may be written step by step." The term "phonograph" was used in the first part of the twentieth century for the mechanical record player that came to be known as a gramophone.

14. *how you came upon this idea*: Lemaître's motivations for proposing his primeval atom hypothesis are not entirely clear and cannot be reconstructed from the extant source material. Various possibilities are discussed in Kragh and Lambert (2007).

15. *what one might call entropic creation*: Briefly, the argument is that since the entropy of the world always increases, and the present entropy is far from its maximum value, the world cannot have existed in an eternity of time. See also Chapters 2 and 4.

16. *time-scale consistent with the age of stars*: In a report of 14 January 1932 to the Committee for Relief in Belgium, Educational Foundation, Lemaître wrote that he had replaced Laplace's hypothesis of a primeval nebula by a primeval atom hypothesis: "An evolution from this starting-point would meet the requirement of the time-scale and furthermore would prove a natural explanation of the penetrating cosmic radiation." (Kragh 1996, p. 50). See also the interview with Willem de Sitter in Chapter 6.

17. *motivated by the existence of radioactive elements*: "The idea of this [primeval atom] hypothesis arose when it was noticed that natural radioactivity is a physical process which disappears gradually and which can, therefore, be expected to have been more important in earlier times." (Lemaître 1949b, p. 452). The idea of radioactivity as a cosmic arrow of time indicating a finite-age universe was first proposed by A. E. Haas in 1911. For references and discussion, see Kragh (2007a).

18. *something that Millikan had proposed*: The American physicist Robert A. Millikan, a Nobel laureate of 1923, argued in the late 1920s that high-frequency starlight formed material particles in the depths of space. Lemaître picked up the idea, which he suggested might have cosmological consequences. "One could concede," he wrote in a paper of 1930, "that the light had been the original state of matter, and that all the matter condensed in the stars was formed by the process proposed by Millikan." (quoted in Kragh and Lambert 2007, p. 456).

19. *Eddington's address on entropy and the beginning of the universe*: A reference to Eddington (1931a), in which he defended the hypothesis of a future heat death caused by the unavoidable growth in entropy. As to the cosmic past he wrote: "Philosophically, the notion of a beginning of the present order of nature is repugnant to me." (Eddington 1931a, p. 453).

20. *You apparently wrote it as a private person*: Lemaître could have signed his letter "G. Lemaître, Catholic University, Louvain," but he chose "G. Lemaître, 40 rue Namur, Louvain."

21. *address to the British Association meeting*: In 1931 Lemaître argued that the cosmic rays were "ashes and smoke of the very rapid fireworks … glimpses of the primeval fireworks of the formation of a star from an atom, coming to us after their long journey through free space." (*Nature* **128** (1931): 705).

22. *photons as some prefer to call them*: The name "photon" was coined by the American physical chemist Gilbert N. Lewis in 1926, although it had been used as early as 1916 as a unit in physiological optics. Lewis' concept of the photon was quite different from Einstein's light quantum for other reasons, because the number of photons satisfied a conservation law. Nonetheless, physicists quickly adopted "photon" as synonymous with a light quantum, such as illustrated by the famous 1927 Solvay Congress on "Electrons and Photons." Throughout the 1930s both names were in use, although some physicists (including Einstein) never used the new term.

23. *Millikan is out*: For an account of the dispute over the nature of the cosmic rays and their cosmological significance, see De Maria and Russo (1989). By 1938, the time of the interview, Millikan's ideas concerning the cosmic rays were no longer being taken seriously.

24. *fortunately I collaborate with Vallarta*: The Mexican physicist Manuel Sandoval Vallarta collaborated in the 1930s with Lemaître. In a paper of 1933 they concluded that their calculations gave "some experimental support to the theory of a super-radioactive origin of the cosmic radiation." (Lemaître and Vallarta 1933, p. 91).

25. *matter as condensed cosmic rays*: According to Lemaître (1949a, p. 366), "All kind of matter must be present in the cosmic rays and matter is nothing else than condensed cosmic rays." At the time nuclei with atomic number up to about 40 had been identified in the penetrating radiation, and he thought that these "super alpha-rays" supported his hypothesis. There is a certain similarity between Lemaître's ideas and Birkeland's much earlier speculations of a space plasma filled with decay products of "electro-radioactivity," such as described in Chapter 1.

26. *a kind of heavy electron*: In the spring of 1937 Carl Anderson and his collaborator Seth Neddermeyer concluded that they had found a charged particle in the cosmic rays with a mass intermediate between the electron and the proton. The discovery of the "mesotron"—today called a muon (μ)—took

physicists by surprise. Anderson was present at the Notre Dame symposium, where he spoke on the new particle and its role in the cosmic rays.

27. *It's just there, for no good reason*: The discovery of the muon caused problems for
physicists, who could not understand the new particle theoretically. Isidor
Rabi, a Nobel laureate of 1944, is to have quipped, "Who ordered that?"

28. *the suggestion that Baade and Zwicky came up with*: Walter Baade and Fritz Zwicky
suggested that ionized gas shells expelled at great speeds from a supernova were the source of cosmic rays, which "should consist of protons and
heavier ions." They further speculated that the end result of the explosion
would be a "neutron star" consisting mainly of densely packed neutrons.
"We realize that this suggestion is highly speculative," they wrote, "in
view of our complete ignorance with respect to the evolution of the universe." (Baade and Zwicky 1934). While the term "neutron star" was coined
by Baade and Zwicky, "supernova" had been introduced by the Swedish
astronomer Knut Lundmark two years earlier.

29. *there is an aesthetic element in the issue*: Lemaître (1949b) contains his main
arguments in favour of the cosmological constant. In a letter of 1947 to
Lemaître, Einstein described the constant as "very ugly" and went on to say:
"About the justification of such feelings concerning logical simplicity it is
difficult to argue. I cannot help to feel it strongly and I am unable to believe
that such an ugly thing should be realised in nature." (quoted in Kragh
1996, p. 54). At about the same time, Lemaître referred in an unpublished
manuscript to Einstein's and others' rejection of the lambda constant as
"prejudices of a psychological and aesthetic origin." See Godart and Heller
(1978, p. 352).

30. *as if empty space has a definite energy density*: On 20 November 1933, while staying
at the Catholic University of America, Washington, DC, Lemaître (1934)
gave an address to the National Academy of Sciences in which he argued
that the cosmological constant Λ corresponds to a vacuum energy density
$\rho = \Lambda c^2/8\pi G$ and vacuum pressure $p = -\rho c^2$. His insight attracted no attention
at all but is today recognized as a partial anticipation of the dark energy discovered more than sixty years later (Kragh 2012b). For more than twenty
years Lemaître's paper received no citations, not even by himself.

31. *I do believe that the universe is finite*: Lemaître later spoke of "the nightmare of
infinite space." He said: "The universe is not too large for man; it exceeds
neither the possibilities of man nor the capacity of the human spirit." See
Godart and Heller (1978, p. 359).

32. *I once discussed the question with Tolman*: Referring to conversations with Lemaître,
Tolman (1934, p. 484) wrote that Lemaître's preference for a closed model
reflected his epistemic optimism, since "an infinite universe could not be
regarded in its totality as an object susceptible to scientific treatment."

33. *Haas, our host here at Notre Dame*: Arthur Erich Haas was an Austrian physicist who mostly worked in atomic and quantum theory. In 1935 he migrated to the United States, where he joined the faculty of Notre Dame University. During his brief career in the United States he mostly dealt with issues of cosmology. A Catholic, he argued for a finite-age closed universe. "Modern science," he asserted, "is getting farther and farther away from the belief in infinite universe." (Haas 1938, p. 5).

34. *what Schwarzschild said a long time ago*: As mentioned in Chapter 3, Schwarzschild preferred for epistemic reasons a closed universe with only a finite number of stars. Such a universe could in principle be fully known to science, whereas an infinite universe could not.

35. *science is about nature and the Bible is about salvation*: "The idea that because they [the writers of the Bible] were right in their doctrine of immortality and salvation they must also be right on all other subjects is simply the fallacy of people who have incomplete understanding of why the Bible was given to us at all." Interview by the journalist and author Duncan Aikman published in the *New York Times Magazine* 19 February 1933. Part of the interview is reproduced in Kragh (2004, pp. 141–152), which offers a summary account of Lemaître's view of science and religion.

36. *that one could understand Genesis better*: In a manuscript of 1921 Lemaître, at the time still a student of physics and theology, used the theory of blackbody radiation to explain the meaning of the first verses of Genesis. The manuscript is reproduced in Stoffel (1996, pp. 107–111).

37. *a universe that agreed with your faith*: The young Dutch-American astronomer Bart Jan Bok arrived at Harvard University in late 1929, where he later met Lemaître. Bok recalled: "LeMaitre [sic] and I were good friends, he said, 'Bart, I've had a funny idea. Maybe the whole universe started out of a single atom, and it exploded, and that's where it all comes from, ha ha ha.' ... Then he came back two years later, and he had, 'I now have a theory of the origin of the universe.' And it was very interesting. The reason why LeMaitre felt so pleased about it: he was a priest ... [and] always said that his universe really fitted his Roman Catholic prejudices and religion because it started with a Big Bang at the beginning, and the good Lord really knew what he was doing." Bok, AIP interview 1978 (see AIP Interviews).

    Although this is an interesting recollection, it is unsupported by independent evidence and cannot be considered reliable. In the period between 1929 and 1933, Lemaître only stayed at Harvard in September 1932, in connection with the fourth General Assembly of the International Astronomical Union.

38. *My theory doesn't support any religious view*: At the 1958 Solvay conference Lemaître stated that his theory of the primeval atom "remains entirely outside any

metaphysical or religious question," and that "It leaves the materialist free
to deny any transcendental Being." (Lemaître 1958b, p. 7).

39. *The universe would still have been created by God*: Christian theology operates with
two different kinds of creation, an original and temporal creation (*creatio
originans*) and a continual and atemporal creation (*creatio continua*). The essence of divine creation is that the world is entirely dependent on God's
will, not that He caused it to exist in the past. See, for example, Kragh (2004,
pp. 242–246). Nonetheless, it is commonly (but mistakenly) believed that
an eternal universe is incompatible with Christian faith. What would be the
consequences if science one day demonstrated that the universe has existed
in an eternity of time? According to the American astronomer and science
popularizer Carl Sagan, "this is the one conceivable finding of science that
could disprove a Creator – because an infinitely old universe would never
have been created." (Sagan 1997, p. 265).

40. *the Creator is essentially hidden*: Concerning the primeval atom theory, Lemaître
(1958b, p. 7) stated that, "It is consonant with the wording of Isaias speaking of the 'Hidden God' hidden in even the beginning of the universe." In
his original manuscript for his 1931 letter to *Nature*, Lemaître ended with
a paragraph that he crossed out: "I think that everyone who believes in
a supreme being supporting every being and every acting, believes also
that God is essentially hidden and may be glad to see how present physics
provides a veil hiding the creation." See Kragh (1996, p. 49).

41. *an American writer interviewed me*: Duncan Aikman, as mentioned above,
and also "Salvation without belief in Jonah's tale," *The Literary Digest* **115**
(11 March 1933): 23.

8

Arthur Eddington's Rationalistic Cosmology

Interview conducted at Professor Eddington's residence in Cambridge Observatory on 2 December 1938. Language: English.

It was on the advice of Haas and Lemaître that I contacted Eddington, who kindly agreed to meet me in Cambridge. I had not left the United States since my arrival four years earlier, but now had business obligations that demanded my presence in the Old World. I went by aeroplane to Paris and from there to Amsterdam and London, where I spent two weeks. I had prepared myself for the meeting with Eddington by reading some of his books and recent papers, which I found to be fascinating but unfortunately also very hard to understand. To my regret, I was forced to conclude that his new theory integrating the quantum and the universe went beyond my comprehension. While in England I also thought of interviewing Arthur Milne in Oxford, but due to various misunderstandings this turned out not to be possible.

The skies over Europe had darkened since I left the continent four years earlier. Not only had Austria lost its independence and become a province of Germany, but as a result of the Munich agreement the Sudetenland in Czechoslovakia had also been occupied by the Germans. I distinctly felt the dense atmosphere and the fear of a new war. Few of those I talked with in England had much confidence in Chamberlain's assurance that the treaty signed in Munich meant peace for our time. But nothing of this entered the conversation I had with Eddington (Figure 9) on that rainy day in peaceful Cambridge.

CCN Let me first express my gratitude that you agreed to share with me some of your thoughts concerning the structure of the universe and its relation to the subatomic world. I imagine that this ambitious project of yours began when you realized nearly ten years ago the significance of Lemaître's paper on the expansion of the universe, and then you ... hmm, you ...

Figure 9 Arthur Stanley Eddington. Photograph by Howard Coster. Source: Douglas (1956, Plate 1).

ASE It actually started a little before that and in a branch of physics that apparently had nothing to do with the universe. Lemaître's paper was important, of course, but the immediate inspiration came from Dirac's relativistic theory of the electron[1] which caused me to think about the meaning of the wave equation and its possible significance for the universe at large.

CCN Really? Well, we need to return to that, but first I'm curious about how you responded to Lemaître's paper of 1927. I understand that he was a former student of yours and that he sent you a copy of the paper?

ASE Yes, before he went to the United States he spent a year in Cambridge, studying relativity theory. He was very bright—I mean *very* bright—and when he returned to Belgium I strongly recommended him for a position.[2]

CCN Were you initially aware of his paper? I have talked with Lemaître and also with de Sitter about the delayed recognition of the expanding universe.

ASE It's a little embarrassing, this episode. You see, what happened was that in early 1930, together with McVittie,[3] I was investigating the stability of the Einstein world; then, a little later I received a letter from Lemaître[4] in which he reminded me about his theory offering an escape from the dilemma between the Einstein and de Sitter universes. Somehow I had forgotten it, I cannot even remember if I read it in 1927 or just put it aside. I must have put it aside, I guess. Anyway, we then immediately realized that it was the right thing.

CCN Yes, and with the endorsement of you and de Sitter the expansion of the universe quickly became accepted by the majority of astronomers. And what about the physicists? More specifically, do you know when Einstein became aware of and accepted the expanding universe?

ASE Oh, I think I had some small share in Einstein's conversion, at least in his first doubts concerning the static model. You see, he stayed with me in the summer of 1930,[5] when we discussed cosmological issues among several other matters. I actually have a nice photograph, where we are sitting in front of my study, almost where we are sitting now. Would you like to see it?

CCN Yes, very much so.

ASE Here it is, it was taken by my sister. Such a beautiful summer, it was. Well, at the time he was not acquainted with the new developments, so I used the opportunity to update him on both the theoretical and observational advances. He was not ready yet to admit that the universe really expands, but I think he was warming up to the idea. While we did not quite agree on the cosmological problem—and especially not on the question of the cosmical constant—we very much agreed that pacifism was more necessary than ever. Apparently we no longer do, but that's another matter.

CCN Yes, we live in a troubled time. Now, I understand that today there are still people who question the expansion and prefer explanations of the redshifts based on a static universe. There are various attempts relying on the hypothesis that the energy of light decreases[6] from the time light is emitted from a star until it's received on Earth. And recently I have come across a couple of papers claiming to explain the Hubble relation on the hypothesis that the speed of light decreases with cosmic time[7]—that a billion years ago light moved much faster than it does today. Could this be possible?

ASE Absolutely not! I know of these amateurish speculations which are not only completely ad hoc but also flatly contradict the theory of relativity. The speed of light in a vacuum cannot *possibly* change—it is a self-contradictory claim, because the speed of light constitutes the very basis of measuring distances. So it's literally nonsense, you can safely forget about it.

CCN Thank you for the advice, I found the hypothesis rather weird myself. But is it only the speed of light that must remain the same, or is this also the case with the other constants of nature? I'm asking because Dirac has recently proposed a cosmological theory[8] based on the assumption that the gravitational constant decreases slowly in time.

ASE Oh, that idea, I'm afraid that it's no less nonsense than the idea of a varying speed of light. I must admit that I am puzzled how a genius like Dirac can come up with such a weird hypothesis. It may be nonsense on a higher plane, but it's nonsense all the same and I prefer to ignore it—as indeed do all astronomers.

CCN Now, you said that Dirac's quantum theory of the electron initiated the line of thinking that has led to your recent ideas of the universe and which you are still working on. The connection is far from obvious, but is it correct that it has something to do with the constants of nature of which the speed of light is but one example? Could you please indicate how you get from the tiny electron to the vast universe?

ASE As you probably know, I have published my views in several articles and books of a popular nature, most recently in *New Pathways in Science*. But now that you have come all the way from America to visit me, let me try to outline some of the considerations that brought me from Dirac's electron to the number of particles in the universe. First I rewrote Dirac's wave equation in a more symmetric way, from which I deduced that the fine structure constant[9] has the precise value of 137—well, I actually first got 136 but soon realized that because two electrons cannot be distinguished it implies an extra unit. Although not itself of cosmological significance, it's an important deduction and the first time that the value of the constant has been determined purely theoretically.

CCN I find it most fascinating that the fine structure constant is a whole number, as if there were some deep connection between fundamental physics and whole numbers. Does your figure 137 agree with what experimenters have found for the individual constants?

ASE It's remarkably close, although some experimental physicists maintain that it disagrees slightly with their measurements. Actually, if one calculates the observed value of the fine structure constant according to my theory one obtains complete agreement between observation and theory. Future measurements might turn out to disagree with my deduction, but even then it wouldn't worry me. It would be a problem for the experimenters and not for my theory.[10]

CCN What? Are you saying that theory should be given priority over experimental facts?

ASE Yes, sometimes theory should be rated higher than *so-called* experimental facts. In any case, I discovered that the atomic constants can be related to constants that characterize the universe as a whole—indeed that the relations follow as necessary consequences from a correct understanding of quantum mechanics and relativity theory. You see, the standard view is that the Dirac equation applies to a single electron, but in my view this is a complete misconception. There are two basic principles of quantum mechanics, the exclusion principle and the uncertainty principle, and they both tell us that when we measure a physical system the entire rest of the universe is involved in the measurement. One must interpret the mass term in Dirac's equation as an energy exchanged with all the other charges in the universe. In this way I obtain a definite relationship between the atomic constants and the total number of electrons in the universe.[11] The atom and the universe dance in harmony.

CCN The total number of electrons in the entire universe! Are you really able to calculate this huge number?

ASE Indeed I am, and I call it the cosmical number. It follows rigorously from my theory, where it comes out close to 10^{79}.[12] Since matter is electrically neutral there must be an equal number of protons. Others of my formulae relate to the ratio of the electric force and the gravitational force, which is nearly the same as the square root of the cosmical number, that is, of the order 10^{39}. You may believe that the number of particles in the universe, the cosmical number, might have been different, but this is not how I look at it. No, it's a true constant of nature and no less fundamental than other constants such as the mass of the electron and the gravitational constant. The cosmical number is what it is because it has to be what it is.

CCN But how? I just don't understand . . .

ASE I think you will if you study my arguments carefully, but of course you'll also have to master the mathematics which is far from trivial. In any case, the connection between atomic theory and cosmology automatically involves the cosmical or cosmological constant, which to me is the cosmic yardstick that fixes the radius of the universe. It's thus indispensable, and I greatly disagree with Einstein and other physicists who take the constant to be zero. It just cannot be, for that would be knocking the bottom out of space.[13] Fortunately I'm not alone in defending the constant. Lemaître considers it no less important than I do.

CCN He certainly does, and I also think that Tolman finds lambda attractive, much like de Sitter did.

ASE Yes, and from my point of view the cosmical number and the lambda constant are connected[14] in the sense that if one knows the cosmical number, one also knows lambda. In fact, I have calculated the cosmical constant in terms of other constants of nature[15] and in this way derived the limiting speed of the recession of nebulae or, what is practically the same, a value of Hubble's constant.[16] I obtain a value not too far from the one Hubble finds observationally. So you see, the atomic world and the universe are really bound inextricably together.[17] When we look inside the atom, we find the keys to unlock the secrets of the cosmos—and vice versa.

CCN I understand from what you say that you conceive the universe as finite and spherical, much like Lemaître does, and that you also share his commitment to the cosmological constant. But do you also agree with him that the universe is of finite age, perhaps with an explosion-like beginning?

ASE No, I do not, emphatically not. I find his original model much more appealing, and this is the kind of model I'm trying to incorporate in my unified theory. The expansion started gently in an asymptotic manner from an Einstein-like equilibrium state of radius a few hundred megaparsecs and a density about a hundred or perhaps a thousand times as high as that of the present universe. It's about one hydrogen atom per litre. As a result of the expansion, the universe has expanded by a factor of ten or so.[18] This is a picture quite different from Lemaître's and much more attractive since it does not require a sudden beginning of things.[19] I don't believe that the universe started in a bang—in that case it must have been an incredibly big bang. So, in my view, although the expansion slowly accelerated, there is no real beginning or birth of the universe.

CCN   This is if we go far back in time, but what if we go into the far future, billions of years from now?

ASE   Well, although the universe doesn't have a birth it will suffer a kind of death, but not in the sense that space contracts into a point. The universe will die a heat death,[20] which is an inevitable consequence of thermodynamics and something we shouldn't bemoan. But before that happens, other and strange things may happen, such as I pointed out several years ago. In the kind of universe with a cosmical constant that I favour—and which Lemaître also favours—the expansion will accelerate. As a result, at some stage the distant nebulae will fly away from us with a speed greater than the speed of light, and this leads to a situation where our universe splits up, as it were, into many distinct universes.[21] We will not be able to see these other universes, not even in principle, but they nonetheless exist.

CCN   Many universes? Nebulae that move faster than light? It becomes more and more strange. . . . Doesn't relativity theory tell us that a body cannot possibly move with a speed greater than that of light?

ASE   Yes, it does, but you need to recall that the nebulae are not moving through space. They are glued to space, so to speak, and it is space that is expanding at an ever greater speed. There's no problem with the postulate of relativity theory.

CCN   Of course, now I see. Just for the record, are you sure that the universe is finite in space?

ASE   Yes, according to my theory it must be finite. Observations cannot give us an answer, but theory can.

CCN   Thank you. Two years ago you published your great book on the relativity theory of protons and electrons, which I have dutifully looked at but far from understood. From what little I understand it seems clear that it is a most ambitious work and also one that is based on a method of science that differs quite drastically from the one usually adopted. You don't seem to have much faith in the empirical method based on experiment and observation?

ASE Not faith, perhaps, but I don't disregard experiments or consider them unnecessary, I just point out that they are of limited importance and especially so when it comes to the *fundamental* aspects of physics, such as the laws and constants of nature. I admit that we cannot establish the chemical composition of a particular mineral without recourse to experiment, but this is not what I call fundamental knowledge. In the same vein, whereas I can calculate the cosmical number purely theoretically, I cannot do the same with Avogadro's number.

CCN Let me see, Avogadro's number, isn't that the one that tells us how many molecules there are in a lump of matter? As a former chemist I ought to know.

ASE Yes, it is, it's a useful but not a fundamental constant. My own theory is solely concerned with fundamental questions, and for that reason it doesn't rest at all on empirical tests and data. It's a purely deductive theory[22] based on epistemological and not empirical considerations. Experiments cannot verify or falsify the deductions any more than measurements with a ruler and a compass can verify or falsify the truths of geometry. Ultimately, if measurements contradict some of the results I have deduced theoretically it must be because the measurements are wrong.

Another way of expressing the methodology behind my theory— and it's one that inevitably and much to my delight provokes my colleagues in physics—is to say that my insights are of an a priori nature. In a very real sense, the laws of nature are imposed by the mind. They are determined by our mental faculties. They are not assertions about or generalizations from experience and consequently they cannot be violated by experience.

CCN This seems to be a radical position indeed, so different from the ordinary way of thinking in physics. I wonder how much it differs from the one expounded by Milne in his cosmological system. You see, I follow the discussions in *Nature*, where there was recently this big debate[23] on the proper methods of science and where your methods and those of Milne were apparently placed under the same hat. Milne too has been accused of rationalism and disregard of observation. Would you agree that there is some similarity?

ASE There may be some superficial similarity in method, but certainly not in substance. Not only is Milne's kinematic cosmology[24] entirely foreign to general relativity, I also find it very artificial—a play with words and symbols. It assumes a uniform expansion from some initial point, but it is merely an assumption based on some speculations concerning measurements of length and time. Relativity theory predicts that the expansion will accelerate,[25] not be uniform. Moreover, his universe is infinite and contrary to mine there is no connection between cosmology and atomic physics. And then he seems to deny the second law of thermodynamics, not to mention that he turns God into an ally of his cosmological system.[26] I must admit that I find it hard to understand why people here in England can find his system so attractive—not only Walker but also McCrea and young Whitrow[27] take it seriously.

CCN I understand that your own theory has met with considerable opposition among both physicists and philosophers, and that it is widely considered obscure or even unintelligible. How do you feel about the reception among scientists?

ASE Well, this is my personal project.[28] I'm working on it pretty much alone and it's far from finished. I am convinced that in due time it will be recognized as a revolution in our understanding of nature, an advance comparable to what happened when Copernicus replaced the Earth with the Sun as the centre of the universe—it sounds terribly immodest, I know that, but what I'm aiming at is nothing less than a revolution. Yes, you're right, there are these objections about obscurity and even mysticism, but they are not fair and to a large extent they are based on superficial and sometimes biased readings of my work. Perhaps it is only natural that new and revolutionary thoughts are initially seen as obscure[29]—that was the original response to Maxwell and Einstein, but eventually people studied them carefully and then the obscurity vanished.

CCN And the same will happen with your theory?

ASE I think so, but it may take some time. I know very well that some of the leaders of quantum physics find my ideas to be entirely wrong, but they have not really understood them, or perhaps they don't *want* to understand them. Earlier this year I attended a meeting in Warsaw[30] where I gave a talk in front of Bohr and several other quantum luminaries belonging to his school. I pointed out that ordinary quantum mechanics fails to take into account the necessary connection to the universe as a whole and that my theory offers a more rational foundation for quantum mechanics than the present one, which in part relies on empirical methods. But it was as if I couldn't communicate with them. Bohr and Kramers came up with objections[31] based on the orthodox understanding of quantum theory, and they were unwilling to consider the possibility that their orthodoxy might be wrong.

CCN Do all quantum physicists share this negative attitude to your theory?

ASE Fortunately not, there are a few with minds open enough to question the Copenhagen orthodoxy and take a genuine interest in my point of view. Schrödinger, the father of wave mechanics, is quite enthusiastic about my approach and wants to develop it. But he had the same experience as I, that Bohr and his people just won't listen. All the same, it is a consolation for me to have an ally in the person of Schrödinger.[32]

CCN I wonder if you don't have more allies. I talked earlier this year with Arthur Haas at Notre Dame University who is greatly interested in your work and has suggested various relations between the constants of nature[33] in accordance with your line of thought.

ASE I know of some of his papers, but they are of a rather numerological kind. They seem to lack the rational justification and foundation in theory that is essential to me. Anyone can play around with numbers, that's not what I'm doing. I am not in the business of either numerology or constantology.[34] But let me add that Lemaître has also taken a keen interest[35] in my work, although in his case mostly in the mathematical aspects.

CCN A propos, in my conversation with Lemaître earlier this year I asked him about the relationship between religion and science. It is well known that you are a religious man, so I wonder if your view of nature, and your view of the universe in particular, if it's in some way influenced by your faith?

ASE No, not really. In a general sense I consider modern physics and cosmology to be compatible with my religious perspective, but the two are different. While the observed world is governed by the laws of physics, the spiritual world is not. As to cosmology, my dislike of a sudden beginning of the universe is consonant with my religious sentiments,[36] but it is certainly not derived from it. That's all I can say.

CCN Thank you. Professor Eddington, before we end our session I would like to have your opinion on what you are primarily known for by the man in the street, or the better educated man in the street, namely, your many popular books about science. They have made you quite famous not only in England but also in the United States and Europe. Is that something you just do for your pleasure, or do you have other reasons to engage in a literary genre that is foreign to most active scientists?

ASE I find it fairly easy to write on this level and apparently have some talent for it, and Cambridge University Press is very eager that I write more of the kind. I find it natural for a scientist, and especially for an astronomer, to disseminate modern knowledge to lay people—in fact, we have always done that, there's really nothing new to it.

CCN I have read most of your books, and invariably with great pleasure. Some of them are very easy to read, but there are also some that are more demanding and relate more to your own current work. Is it also a way of promoting your theories?

ASE I guess it is, but I'm also writing them for my own sake, not only for the sake of the readers. What I mean is that when one writes about difficult things for readers with no scientific training, one is forced to present them in a simple and lucid way, and basically with no mathematics. And that's a very useful exercise, for then the essence of the theory comes through better. I myself get a clearer grasp of what I'm thinking and doing. I am currently preparing a new book to be published next year,[37] and although it is popular in a certain sense it is also a serious and quite detailed account of my ideas about the nature of science. There is no need to distinguish sharply between scientific publications and those of a more general and perhaps popular kind. At least, I don't.

CCN Thank you, I look forward reading your new book. Well, your thoughts on cosmology and its possible connections to the quantum world have been most stimulating. My sincere thanks for having shared them with me, it's really been a privilege.

Notes

Arthur Stanley Eddington was born in Kendal, England, on 28 December 1882 and died in Cambridge, England, on 22 November 1944. After studying in Manchester, Eddington entered Trinity College, Cambridge, from where he graduated in 1909. Following a period as assistant at the Royal Observatory, Greenwich, he was appointed in 1913 as Plumian Professor of Astronomy and Experimental Philosophy at Cambridge University and director of the Cambridge Observatory (see Figure 9). Focusing on the interior structure of stars and the sources of stellar energy, Eddington soon established himself as a leading astrophysicist. This line of work culminated in 1926 with the publication of *The Internal Constitution of the Stars*. He suggested two nuclear processes as the source of the energy emitted by the stars, proton–electron annihilation and the transformation of four protons into a helium nucleus. He also suggested a mechanism that explained the variable luminosity of Cepheid stars. As a Quaker and convinced pacifist he avoided military service in World War I without being jailed. When he received news of Einstein's theory of general relativity in 1916, he quickly turned himself into a specialist in and supporter of the new theory and its astronomical consequences. Together with Frank Dyson, the Astronomer Royal, he organized the famous 1919 solar eclipse expedition that confirmed Einstein's prediction of the deflection of starlight by the Sun's gravitational field.

Eddington's mastery of relativity theory and his talent for writing popular works on difficult subjects made him a celebrity in the interwar period, when he was Britain's most well-known scientist. He followed the development of relativistic models of the universe, and in 1930 he enthusiastically supported the expanding model suggested by his former student Lemaître. On the other hand, he rejected the idea that the universe had a beginning. From 1929 until his death in 1944 he developed an ambitious project of unifying quantum mechanics with cosmology, which he described in an enigmatic book of 1936, *Relativity Theory of Protons and Electrons*. His efforts to revolutionize fundamental physics were met with universal scepticism. The main result, *Fundamental Theory* (Eddington 1946), published after his death, was a posthumous failed theory of everything.

Biographical sources: Douglas (1956); Stanley (2007); Dingle (1945). Douglas' biography includes a list of Eddington's published works.

1. *Dirac's relativistic theory of the electron*: In early 1928 Paul Dirac published a new wave equation of the electron which, contrary to Schrödinger's, satisfied the requirements of relativity theory and explained the spin. According to Dirac's theory, an electron is characterized by four wave functions (ψ_1, ψ_2, ψ_3, ψ_4), two of which refer to the spin states of the electron. In 1931 he interpreted the other two wave functions as belonging to a hypothetical "antielectron," a positive electron that soon became known as the positron.

2. *I strongly recommended him for a position*: In a letter of 24 December 1924 to Théodore de Donder in Brussels, Eddington wrote: "I found M. Lemaître a very brilliant student, wonderfully quick and clear-sighted, and of great mathematical ability. ... In case his name is considered for any post in Belgium I would be able to give him my strongest recommendations." (quoted in Douglas 1956, p. 111).

3. *together with McVittie*: George Cunliffe McVittie completed his doctoral dissertation on unified field theories under Eddington in 1930. On the instigation of Eddington he investigated the effect of perturbations in a static universe and later also in an expanding universe. Following his stay in Cambridge, McVittie embarked on a distinguished career in astronomy and cosmology. In 1952 he moved to Urbana, USA, to take up a position as professor of astronomy at the University of Illinois. On his retirement in 1972, he returned to England.

4. *I received a letter from Lemaître*: For Lemaître's letter to Eddington, see Nussbaumer and Bieri (2009, p. 123) McVittie (1967, p. 295) later recalled "the day when Eddington, rather shamefacedly, showed me a letter from Lemaître which reminded Eddington of the solution to the problem which Lemaître had already given. Eddington confessed that although he had seen Lemaître's paper in 1927 he had forgotten completely about it until that moment."

5. *he stayed with me in the summer of 1930*: Einstein was in England to receive an honorary doctoral degree from Cambridge University. For his stay with Eddington in June 1930, see Nussbaumer (2014) and Douglas (1956, p. 102). The photograph taken by Winifred Eddington is reproduced in both sources. Although there is no documentation that the two scientists discussed issues of cosmology, it is hard to believe that they did not. At the time Einstein and Eddington were both committed pacifists, if in different ways. Under the impact of German rearmament and the threat of a new European war, from about 1934 Einstein abandoned his pacifist position. Eddington did not.

6. *the hypothesis that the energy of light decreases*: During the 1930s several astronomers proposed "tired light" hypotheses as an alternative to the expanding universe. This class of hypotheses assumes that photons from the distant nebulae lose energy during their journeys through space. Because the energy E of a photon is given by its wavelength λ ($E = hc/\lambda$), the result will be a shift to the red end of the spectrum. By means of a suitable hypothesis, the redshift can be brought into accordance with Hubble's law.

7. *the speed of light decreases with cosmic time*: On these hypotheses and their modern reincarnations in the form of so-called VSL theories (varying speed of light), see Kragh (2011, pp. 185–189). In his posthumously published *Fundamental Theory*, Eddington (1946, p. 8) dismissed the hypothesis of a varying speed of light as "nonsensical" and "self-contradictory," because "it follows from the definition of the ultimate standards of length and time that the velocity of light is constant everywhere and everywhen."

8. *Dirac has recently proposed a cosmological theory*: Dirac presented a full version of his varying-G theory in 1938. See the interview in Chapter 12. Eddington (1939b, p. 234) dismissed Dirac's theory as "unnecessarily complicated and fantastic."

9. *the fine structure constant*: This dimensionless constant is a relative measure of the strength of electromagnetic interactions. It is given by $\alpha = 2\pi e^2/hc$, where e is the elementary charge, h Planck's constant, and c the velocity of light. Eddington always used the name for the inverse quantity, that is, he took it to be $\alpha^{-1} = hc/2\pi e^2$. Because it is a pure number, its value of about 137.036 is independent of the system of units. For details and literature see Kragh (2003). General descriptions of Eddington's ambitious attempt to unify cosmology and quantum mechanics can be found in, for example, Singh (1970, pp. 239–253) and Kragh (2011, pp. 92–101).

10. *it would be a problem for the experimenters and not for my theory*: Referring to the experimental value $\alpha^{-1} = 137.009$, this is what Eddington (1944a) concluded in one of his last papers. He generally rated theory higher than experiment. "I hope," he wrote (Eddington 1935, p. 211), "I shall not shock the experimental physicists too much if I add that it is also a good rule not to put overmuch confidence in the observational results that are put forward until they have been confirmed by theory." He espoused the same anti-empiricist message in his popular book on the expansion of the universe (Eddington 1933, p. 17), in which he wrote: "For the reader resolved to eschew theory and admit only definite observational facts, all astronomical books are banned. There are no purely observational facts about the heavenly bodies."

11. *the atomic constants and the total number of electrons in the universe*: Among Eddington's relations was $e^2/GmM = 2\pi/\sqrt{N}$, where N is the "cosmical number" or the number of electrons in the universe; m and M refer to the mass of the electron and the proton, respectively. The ratio on the left-hand side signifies the ratio between the electrical force and the gravitational force between an electron and a proton. As early as 1919 the German mathematician Hermann Weyl had suggested a cosmological interpretation of the quantity e^2/GmM.

12. *it comes out close to 10^{79}*: Eddington deduced $N = 2 \times 136 \times 2^{256}$ or approximately $N = 3.15 \times 10^{79}$. In Eddington (1939c, p. 170) he expressed it more dramatically: "I believe there are 15,747,724,136,275,002,577,605,653,961,181, 555,468,044,717,914,527,116,231,525,076, 185,631,031,296 protons in the universe, and the same number of electrons."

13. *knocking the bottom out of space*: This was Eddington's expression in his widely sold *The Expanding Universe*, an expanded version of a public lecture he gave at a meeting of the International Astronomical Union at Cambridge, Massachusetts, in September 1932. Eddington also wrote, "I would as soon think of reverting to Newtonian theory as of dropping the cosmical constant." (Eddington 1933, pp. 104 and 24).

14. *The cosmical number and the lambda constant are connected*: Eddington (1937, p. 6) called the cosmical number N "a deputy cosmical constant." He usually referred to Λ as the cosmical rather than the cosmological constant.

15. *the cosmical constant in terms of other constants of nature*: For example, one of Eddington's calculations resulted in $\Lambda = (2GM/\pi)^2(mc/e^2)^4 = 9.8 \times 10^{-55}$ cm^{-2}. This value was reasonable but could not be compared with measurements. At the time there were no observational bounds on the constant, except that it had to be very small.

16. *a value of Hubble's constant*: Eddington's theoretical values for the Hubble constant changed as his theory progressed. In a lecture given at Harvard University in 1936 he obtained $H = 432$ km/s/Mpc, which he confirmed two years later in an address in Warsaw. His final calculation resulted in $H = 585$ km/s/Mpc, corresponding to a Hubble time as small as $T = 1.54$ billion years. The sources for the quoted values are Eddington (1937, p. 7), Eddington (1939a, p. 187), and Eddington (1946, p. 10), respectively.

17. *the atomic world and the universe are really bound inextricably together*: As he phrased it in *New Pathways in Science* (Eddington 1935, p. 227): "If we invert the relation of the electron to the universe, we obtain the relation of the universe to the electron. We have only to take this equation describing the electron with the universe as comparison object, and view it, as it were, through the wrong end of the telescope, to obtain the equation describing the universe with the electron as comparison object."

18. *the universe has expanded by a factor of ten or so*: "We conclude that the present radius of the universe is between 1000 and 2500 megaparsecs, and most probably about 1500 megaparsecs." (Eddington 1944b, p. 203).

19. *it does not require a sudden beginning of things*: Eddington's dismissal of cosmo-logical models with a beginning in time was philosophically based. In his Gifford Lectures of 1927 he said: "As a scientist I simply do not believe that the present order of things started off with a bang; unscientifically I feel equally unwilling to accept discontinuity in the divine nature." (Eddington 1928, p. 85). And four years later, shortly before Lemaître proposed his primeval atom hypothesis: "Philosophically, the notion of a beginning of the present order of Nature is repugnant to me." (Eddington 1931a, p. 450). In 1933 he advocated his own "placid theory" over Lemaître's. It seems, he said, "that the most satisfactory theory would be one which made the beginning not too unaesthetically abrupt." (Eddington 1933, p. 56).

20. *The universe will die a heat death*: Eddington was convinced that the increase in entropy would eventually result in a dead universe, but he did not accept the inference from the entropy law to a beginning of the universe. See, for example, Eddington (1931a).

21. *our universe splits up, as it were, into many distinct universes*: Eddington (1931b, p. 415) said: "Objects separating faster than the velocity of light are cut off from any causal influence on one another, so that in time the universe will become virtually a number of disconnected universes no longer bearing any physical relation to one another." Without using the term, he introduced what in the twenty-first century became known as the "multiverse." Contrary to

other kinds of multiverses, the distinct universes in Eddington's scenario were located in the same cosmic space and subject to the same laws of physics. On multiverse theories, see Kragh (2011, pp. 255–290).

22. *It's a purely deductive theory*: "It should be possible to judge whether the mathematical treatment and solutions are correct, without turning up the answer in the book of nature," Eddington (1936, pp. 3–4) wrote in *Relativity Theory of Protons and Electrons*. "My task is to show that our theoretical resources are sufficient and our methods powerful enough to calculate the constants exactly—so that the observational test will be the same kind of perfunctory verification that we apply sometimes to theorems in geometry." On Eddington's philosophy of science see Singh (1970) and Kragh (2004, pp. 162–170).

23. *this big debate*: Instigated by an attack on the "modern Aristotelians" by Herbert Dingle (1937), an astrophysicist and philosopher of science, *Nature* published a series of arguments relating to "Physical Science and Philosophy." See *Nature* **139** (1937): 1000–1010. Among the discussants were Milne, Eddington, Dirac, Dingle, and others from Britain's scientific elite. The debate is described in Kragh (2004, pp. 185–189). Robertson (1936) said about "Milne's weird cosmological theory" and "Eddingtons's unholy union of quantum and relativity theories" that they represented "a return to the outmoded Aristotelian method in science – the attempt to derive natural laws from a priori principles."

24. *Milne's kinematic cosmology*: From a methodological point of view there was more similarity between the theories of Eddington and Milne than Eddington wanted to admit. Both of the two theories or systems were based on a priori principles from which the laws of physics were deduced by rational reasoning and with little or no regard for experimental facts. Eddington (1939b, p. 230) sharply criticized Milne's so-called theory of kinematic relativity, which he considered "perverted from the start."

25. *Relativity theory predicts that the expansion will accelerate*: This is what Eddington (1939b, p. 231) said on some occasions, although he knew that the acceleration is an effect of the cosmological constant. For example, in the Einstein–de Sitter model with $\Lambda = 0$ the expansion decelerates. Milne's uniformly expanding universe was not governed by the general theory of relativity, but the model corresponds to a special solution of the cosmological field equations. What in modern cosmology is known as the Milne model is, however, incompatible with observation.

26. *he turns God into an ally of his cosmological system*: In his monumental *Relativity, Gravitation and World-Structure* of 1935, Milne included a section on "Creation and Deity" in which he argued for the necessity for a creative God. In later works, and especially in the posthumously published *Modern Cosmology and the Christian Idea of God* (1952), he explicitly drew connections between God

and his own cosmological theory. The deeply religious Eddington, who was a Quaker, found such apologetic use of science unacceptable.

27. *Walker but also McCrea and young Whitrow*: Arthur G. Walker, William H. McCrea, and Gerald J. Whitrow were distinguished British mathematicians and physicists who made important contributions to theoretical cosmology. They were inspired by Milne's ideas, although without accepting them uncritically.

28. *this is my personal project*: Eddington developed his research programme in splendid isolation, with no students or collaborators. The closed nature of his research is illustrated by the references in the 14 research papers he wrote on the unified theory between 1929 and 1939. Whereas the average self-reference ratio in physics papers in the period was about 10 per cent, no fewer than 70 per cent of Eddington's references were to his own works.

29. *new and revolutionary thoughts are initially seen as obscure*: In a letter to Herbert Dingle in 1944, the year of his death, Eddington wrote: "I am continually trying to find out why people find the procedure obscure. But I would point out that even Einstein was once considered obscure, and hundreds of people have thought it necessary to explain him. I cannot seriously believe that I ever attain the obscurity that Dirac does. But in the case of Einstein and Dirac people have thought it worth while to penetrate the obscurity. I believe they will understand me all right when they realize that they have got to do so." (quoted in Dingle 1945, p. 247).

30. *I attended a meeting in Warsaw*: The Institute of Intellectual Co-Operation, established in 1924 with the help of the French government, held a conference on "New Theories in Physics" in Warsaw from 30 May to 3 June 1938 (IICO 1939). Among the invited speakers were Niels Bohr, Louis de Broglie, George Gamow, Oskar Klein, Hendrik Kramers, and Eddington. After the Warsaw conference ended, the group spent a couple of days in Cracow. See the photograph in Chernin (1994, p. 800).

31. *Bohr and Kramers came up with objections*: None of the physicists gathered in Warsaw accepted Eddington's ideas. As the Dutch physicist Hendrik Kramers put it: "When listening to Professor Eddington's interesting paper, I had the impression that it concerned another quantum theory, in which we do not find the formulae ordinarily used, but where we find many things in contradiction with the ordinary theory, as Professor Eddington himself realizes." (IICO 1939, p. 194). Eddington replied that he had attempted to apply the same kind of criticism to quantum theory which Einstein had applied to classical physics.

32. *an ally in the person of Schrödinger*: From about 1937 to 1940 Erwin Schrödinger warmly supported Eddington's theory. On the other hand, he also found its formulations difficult to understand, such as he mentioned in a long letter to Eddington of 23 October 1937: "I have just returned from Bologna,

where I endeavoured to give a brief report on those parts of your theory which seem the most important to me. I met with an unvanquishable incredulity of the important group Bohr, Heisenberg, Pauli and their followers. It was an extremely difficult position—spiritually I mean—because so many of your arguments are as ununderstandable to me as they are to them. . . . My suspicion is, that there exist a few very important points, which you explain orderly in the right place, but for some reason or other we misinterpret your words just as if they were Chinese." Schrödinger ended his letter by assuring Eddington that, "there is nothing in the world of knowledge in this moment that interests me more." The letter is reproduced in Meyenn (2011, pp. 591–594).

33. *various relations between the constants of nature*: For example, in 1936 Haas proposed that the net energy of the universe was zero. From this he deduced that the mass μ of the closed universe was given by the simple formula $\mu = Rc^2/G$, where R is the radius of the universe (see Kragh 2004, p. 191). Since the 1970s some cosmologists, attempting to explain the origin of the universe from a zero-energy state, have made use of the same idea.

34. *constantology*: The word first appears in the literature in 1961, coined by the American physicist Wolfgang Yourgrau who used it to ridicule the kind of numerology based on the constants of nature that had its roots in the Eddington epigones of the 1930s (see Kragh 2003).

35. *Lemaître has also taken a keen interest*: Lemaître carefully read the proofs of *Relativity Theory of Protons and Electrons* (Eddington 1936, p. vi). For his critical interest in Eddington's theory, see Lambert (2000, pp. 173–185).

36. *consonant with my religious sentiments*: As a Quaker, Eddington was convinced that the world of science was different from and smaller than the more significant spiritual world (see Stanley 2007; Kragh 2004, pp. 103–112). His view of physics and cosmology was largely unconnected to his religion. However, it has been suggested that his dislike of a big bang universe reflected the conception of God characteristic of the Quakers (Batten 1994) This is also what Bart Bok said in the interview mentioned in Chapter 7. According to Bok, "In Eddington's [universe], everything is smooth and even, and then there's a little ripple of density—a Quaker gets up and makes a few remarks, oops, dies down again, and then slowly, the Eddington universe develops."

37. *a new book to be published next year*: The book was *The Philosophy of Physical Science* (Eddington 1939c). It was the most elaborate exposition of Eddington's mature thoughts and the one that attracted most philosophical interest.

9

Edwin Hubble, Observational Cosmologist

Interview conducted at Hubble's house on Woodstock Road, San Marino, California, on 8 December 1951. Language: English.

In 1942 I attained United States citizenship, and the following year I was requested to join a development project at Union Carbide, the large chemical company. The group of engineers and scientists I became part of mostly worked on methods to separate gases by means of diffusion processes. It was only later that I realized that the work was a small cog in the enormous Manhattan Project that in July 1945 led to the first atomic bomb and subsequently to the bombing of Hiroshima and Nagasaki. This was my first and only contribution to the war efforts of the United States, and not a heroic one at all.

It took a couple of years after the war for me to regain my appetite for cosmology, inspired in part by the completion of the Mount Palomar telescopes and in part by new and exciting theories of the universe such as the steady-state model and Gamow's explosion model. Although I had lived in the United States for a long time, I had never been to southern California. When I got the chance to interview Edwin Hubble (Figure 10), a legend in astronomy and cosmology, I decided to go by car, a new and spacious Oldsmobile 98, and to combine the interview with an extended tour around the south-western states. In 1949 Hubble had had a serious heart attack, but fortunately he had recovered. When I met him and his wife Grace in their beautiful home he appeared to be in good shape.

CCN I wonder if we could begin, just for a moment, by going back in time more than two decades to the late 1920s when you did your celebrated work that led to the recognition of the expansion of the universe.

Figure 10 Edwin Hubble at about the time of the interview with CNN. Courtesy of the Archives, California Institute of Technology.

EH That's fine with me. I wasn't aware then that I had discovered the expanding universe, but that's what people keep telling me—Grace is one of them.[1] Anyway, go on with your questions and let me know if you want a glass of lemonade. It's warm today.

CCN Thank you, I'm not thirsty. I would like to know when you first became aware of the mathematical physicists' discussion of cosmological models based on Einstein's theory of general relativity. I mean, most observational astronomers would have no reason to pay attention to these mathematical and highly abstract discussions—and if they did, they might not have understood them. Am I right?

EH Yes, those theories are abstract indeed, but I had the good fortune of having contact with people with the necessary insight, Tolman in particular; and I tried to follow the development by reading review articles and other secondary literature. I have to admit that I'm not much of a mathematician myself.

When I wrote my paper of 1926 on the classification of the extragalactic nebulae[2] I used the occasion to estimate the mean density of the universe. I knew at the time of the two static models, although I had not read the technical papers of either Einstein or de Sitter. It's a long time ago . . . I think I used formulae from a textbook[3] to transform the estimated density to the size of the Einstein universe. I no longer recall what book it was, but perhaps I said so in the paper— you can look it up. I also had conversations with Tolman and Robertson,[4] but that may have been a couple of years later. My memory is a bit hazy.

CCN Do you recall what the conversations were about? Did they relate to the theoretical predictions of a relationship between redshifts and distances?

EH Sorry, I don't remember. Perhaps Tolman mentioned it to me, but I'm not sure when it was.

CCN Of course, after all, it was a quarter of a century ago. By the way, I have noticed that you always speak of "extragalactic nebulae," whereas many astronomers nowadays just speak of "galaxies." Is there a difference?

EH No, not really. It's just that I find "extragalactic nebulae" a more appropriate name. I introduced it in 1926 and see no reason to change to the new and possibly more fashionable name.

CCN Yes, perhaps it's just a fashion. I assume that you were unaware of the work of Lemaître when you wrote your famous paper in 1929? Most people seem to have been unaware of it.

EH That's true, I only became aware of it the following year, after Eddington and others had made the connection. But I actually met him in 1924 or thereabouts,[5] he did some work under Shapley and visited me at Mount Wilson. Of course, at the time I had no idea that he would soon come up with the idea of the expanding universe.

CCN But in the eyes of the public it's you rather than Lemaître who's recognized as the father of the expanding universe. After all, his work was theoretical and it was you who actually proved the expansion. Am I right?

EH Hmm, both yes and no. My work of 1929 was important because it demonstrated for the first time the law of redshifts, but I did not actually conclude that the universe is expanding;[6] and for that matter, I still feel it is premature to see in the redshift law a proof that the universe is in a state of expansion. Back in 1929 I suggested that the data might represent what at the time was known as the de Sitter effect,[7] not that they were due to the nebulae flying away from us. I did speak of the radial velocities of the nebulae, but what I meant were apparent velocities,[8] the redshifts expressed by means of Doppler's old formula. What matters is that I had established a linear correlation between observational data—redshifts and distances— first somewhat approximate but soon, in the work with Humason,[9] in a much more precise and convincing way. Whatever the theoretical interpretation, the law of redshifts has been firmly established as a fact, and that's the important thing.

CCN I noticed that in some of the recent literature the velocity– distance relation is now called the "Hubble law"[10] and the recession constant the "Hubble constant." Likewise, the velocity–distance graph you presented in 1929 is known as the "Hubble graph." You must be honoured to have a constant and a law of nature named after you, to enter the select company of people like Newton and Planck?

EH Well, yes, I guess I should be honoured, but after all a name is just a name. Wasn't it Shakespeare who said that a rose would smell as sweet[11] even if we were to give it a different name? He's saying "What's in a name" and then. . . . No, I can't remember. Do you remember the quote?

CCN No, I don't. Perhaps we should let Shakespeare rest in peace.

EH We'll let him rest in peace, then. Myself, I prefer to speak of the law of redshifts. On the other hand, it is not unreasonable to associate my name with the constant, and to designate it by the letter H,[12] since I was the first to determine it by means of observations. A little later de Sitter used our data to calculate the constant, unfortunately without giving proper credit to my earlier formulation[13] of the relation and its origin in the Mount Wilson observational programme. Well, having thought of it he readily accepted our priority.

CCN As an empirical relationship the redshift–distance law has long ceased to be controversial. It is one of the very few facts of direct cosmological significance and one that any theory of the universe must comply with; but there are different ideas of how to understand it theoretically and what conclusions to draw from it. Quite recently a group of physicists in England suggested that although the universe expands, its average density remains the same, so there is no real evolution of the universe as a whole. The theory has caused a lot of discussion in England, I understand, but so far it seems to be little known in America. Are you aware of it?

EH I have heard of it, yes, but not studied it. It's my impression that it is not taken seriously by my colleagues in astronomy and physics.[14] I see no reason to dismiss it just because it postulates a continual creation of matter, but the kind of universe it offers may turn out to disagree with measurements—it yields a definite expansion rate which it should be possible to test. There are also the recent observations by Stebbins and Whitford[15] of a systematic reddening of certain nebulae, and they seem to provide evidence of evolutionary changes that cannot be explained by the so-called steady-state theory of Foyle and his colleagues, and yet . . .

CCN Sorry to interrupt, but his name is Hoyle, Fred Hoyle, not Foyle.

EH Yes, right, Hoyle. I think he stayed at Caltech but I don't remember having met him.[16] Well, I just wanted to say that I have no preconceptions against this theory of the steady universe, as some people seem to, but I'm not in favour of it either. It's the privilege of the observational astronomer that he can afford the luxury of staying neutral in theoretical controversies, and it's a privilege I value. So long as observations don't speak out clearly against a particular model or theory one should keep an open mind. I even had some sympathy for Milne's universe,[17] which in a certain sense combined features of the static and the expanding universe. But now, after his death, his theory seems to have been largely abandoned.

CCN I guess you're right. That's my impression too. In your book of 1936, I mean *The Realm of the Universe*, you end with some rather sceptical remarks concerning theoretical cosmology, the last sentence being—yes, here it is—"Not until the empirical resources are exhausted, need we pass on to the dreamy realms of speculation." Is it still your opinion that theories of the universe belong to the dreamy realms of speculation?[18]

EH Well, when I used this expression it was not to denigrate theoretical cosmology in general, only a warning against those parts that cannot be tested observationally. I think I made that clear in the beginning of the book. And this is not an empty warning, for there clearly are cosmological theories that have practically no connection to observation and are in this sense speculations. They may be of mathematical interest, but that's all, as far as I can tell.

I subscribe to what I think philosophers call the empirical-inductive method, just as most of my colleagues do, at least here in America. I'm not saying that speculations have no place in science, only that they need to be restricted to inspirations for ideas that can be turned into testable theories. This is more or less what I have said on various occasions, most recently in a lecture I gave earlier this year.[19] The point is that only observational results can be stated positively and with such accuracy that they rule out one or more theories. I mean, what else should rule them out?

CCN Eddington had some very different ideas on the relative merits of observation and theory.

EH Yes, I know, but that sort of proves my point. After all, his theory remained a mathematical castle in the air, didn't it? Theoretical and mathematical cosmologists deal with possible worlds,[20] but the observer studies the only real world, the one that we live in, and by means of increasingly more accurate and reliable observations he is able to reduce the number of possible worlds to a minimum, hopefully to just a single one. In my view, this is the game of modern cosmology, and one shared by the majority of workers in the field. We have come some way, but as yet far from all the way. There are still too many competing theories and too few observational tests to discriminate between them, but I'm confident that we are heading in the right direction.

CCN This is very interesting. Have you heard of what Hoyle refers to as the "big bang"?

EH The big bang? What a funny name. No, I haven't. Do you mean the Englishman Hoyle who's advocating the theory of continual creation?

CCN The very same. He coined the name for the class of theories he dislikes, namely those that assume a beginning of the universe in a kind of explosion a finite time ago. One of them is Lemaître's, and I'm just curious what you think of his idea—Lemaître's I mean. Would you classify it as a dreamy speculation or one that can be subjected to observational tests?

EH It's not a speculation, although it includes speculative elements such as the primeval atom, but it results in a definite picture of the universe that can be compared with observations, so it's one of several models that merit attention. On the other hand, Lemaître's model fares rather badly[21] when compared with the extensive observations I discussed in the 1930s and which still are valid. I concluded, if I remember correctly, that a closed universe with a cosmological constant of the type proposed by Lemaître must be suspiciously small and of an unacceptably high density, about 10^{-26} g/cm^3 or so, which is several orders of magnitude too high. I'm not particularly disturbed by its explosive origin in a supercompact mass—what Hoyle apparently calls a big bang—but it just seems unlikely that the model universe can be made to agree with observational data. I have no confidence in it.

CCN To the public, cosmology is often associated with human values and religious sentiments, and there is also a tradition among scientists to speculate about the role of man in the cosmos, or about its purpose and things like that. Is that a kind of thinking that appeals to you?

EH No, not at all, that's something I don't want to get involved in. There are people here in California who fancy ideas of this kind and I often receive letters from them—sometimes letters that are decidedly weird. Cosmology is basically extragalactic astronomy, a science like other sciences, and human values have no place in science. Science needs to be objective, and values are not. They are subjective. I sometimes discussed the issue with my friend Tolman, who sadly is no longer with us. He agreed with me that values of a human or religious nature[22] should be kept out of science—there really is no more to say about it.

CCN Even if science and religion should be kept separate, as most scientists believe, then there will always be religious men of a different mind and they will sometimes use the authority of science for their own purpose. In fact, I understand that very recently you have been cited by none other than the pope in Rome[23] and that . . .

EH Oh, that, yes, I was informed of it. I don't take it seriously at all, but there's nothing I can do about it. Just between you and me, let me read from a letter that an old friend of mine sent me[24] two weeks ago.

It's quite funny, here it is, he says: "I am used to seeing you earn new and even higher distinctions; but till I read this morning's paper I had not dreamed that the Pope would have to fall back on you for proof of the existence of God." And then he says, "This ought to qualify you, in due course, for sainthood." That's good, isn't it? Saint Edwin!

CCN That's funny indeed. Thanks for sharing it with me. Now, let's turn to the prospects that the new monster telescope offers for cosmological research. It seems that with this apparatus observational cosmology has almost become a Californian monopoly[25]—or would that be an exaggeration?

EH Not really, it's not much of an exaggeration. There is no doubt that the 200-inch Hale reflector on Mount Palomar is a unique instrument and the only one in the world that can see so deeply into space that it promises to answer some of the big questions about the universe. It penetrates into space about twice as far as the 100-inch[26] and we expect to be able to find redshifts corresponding to at least one-quarter of the velocity of light. There's nothing like it when it comes to exploring the universe and there'll be nothing like it for decades to come. It's a monster, but a beautiful monster.

CCN Could one imagine that the new radio telescopes[27] they are building in England will one day be of use to cosmologists, that radio waves will be as important as the optical waves used in traditional telescopes?

EH I doubt that very much. Radio waves may be important to certain aspects of astronomy, but mostly in the solar system. The people at Harvard consider it a new revelation. But I can't imagine that radio waves will be of any value on the extragalactic or cosmological scale. On the whole I'm not happy with the entrance of radio physics and engineering into astronomy,[28] or for that matter with the new and fashionable trend for nuclear astrophysics. It is as if we astronomers— the real astronomers—are no longer the leaders of our own science, which is of course absurd. Do you know who's in charge of our two Californian observatories?[29] A physicist, not an astronomer!

CCN I see. Well, things are changing. The Palomar telescopes have been under way for a long time, but I understand that now they have become operational. Are you happy with what they have revealed so far?

EH Yes, we have waited for it a very long time. You know, Hale started the project[30] as long ago as the late 1920s, and there were all kinds of problems—one of them a world war! But finally, in early 1949 the 200-inch saw first light. Oh, it was a great day when I targeted a variable nebula with it. It took nearly a year until it became available for research, though, so it's still very new and hasn't yet resulted in dramatic results. Our aim is to obtain better values for the mean density of matter in space and also for the rate of increase of the redshifts, and to find out whether or not there are systematic departures from homogeneity in the distribution of nebulae or from linearity in the law of redshifts.

CCN Could you elaborate, please? I don't quite understand.

EH Yes, you see, by measuring redshifts and magnitudes at very large distances it should be possible to establish whether or not the linear form of the law of redshifts is in need of correction and in this way to get information about the architecture of the universe. Observations with the Palomar telescopes will be able to tell us if the apparent expansion is slowing down—if there is a deceleration[31] or perhaps an acceleration instead. This was a question we considered in the 1930s, but now there's a much better chance of determining the sign and degree of deceleration. I have talked with Sandage about it[32] and he is prepared to look more closely at it—he's much younger than I am and a most promising astronomer. All this will be difficult, I know, but it can be done and then we will be able to distinguish between some of the cosmological theories. But don't expect results to be announced tomorrow. It is a long-term project and one I may not myself see the end of.

CCN The expanding universe has been generally accepted for nearly two decades and yet it has never been definitely proven, as far as I know. What is the verdict based on the data provided by the big telescopes? Does the universe expand or not?

EH You shouldn't expect a simple answer to a difficult question, that's what journalists do, not scientists. At least I cannot offer one. There have been times when I tended to support the standard view that the redshifts are velocity shifts, and other times when I thought that the expanding universe was not vindicated by available data.[33] In the extensive work that Tolman and I did[34] in the 1930s we developed methods for discriminating between possible models of the universe.

We approached the problem with open minds and were forced to conclude that our data didn't even allow us to choose between recessional and static models. Our efforts were not in vain, though, for they gave us rules for deciding observationally between rival cosmological models. The problem was insufficient data.

CCN But hasn't it changed since then?

EH Not really, although in the course of a few years it will. So far there is not, to my mind, an unambiguous answer to the question you posed. The hypothesis that the redshifts are Doppler or velocity shifts is simple and explicable in terms of the laws of physics. If I were a betting man I would probably hold a small sum of money on this kind of answer, but only a few bucks. There are alternative explanations, for instance of the tired-light type preferred by Zwicky and some other physicists,[35] yet to my mind they are doubtful. Now it doesn't mean that expansion is the correct explanation, for it leads to problems both with regard to the density and the age of the universe. The possibility that the redshifts are due to some hitherto unknown principle in nature[36] remains.

CCN What do you mean?

EH Well, some principle that explains the redshifts as something other than due to the Doppler effect. However, it's problematic for methodological reasons.[37] After all, it has the character of an ad hoc explanation, meaning that it's not really an explanation. I don't think I can be more specific. For the moment we need to adopt an agnostic attitude to the old question of whether the universe expands or not. We need to be patient for a few more years.

CCN Thank you, Professor Hubble, it will be exciting to follow the cosmic secrets that Mount Palomar will reveal to us in the future.

Notes

Edwin Powell Hubble was born in Marshfield, Missouri, on 20 November 1889 and died in San Marino, California, on 29 September 1953. After studies in mathematics and astronomy at the University of Chicago he went to Oxford, England, on a Rhodes Scholar stipend. At Oxford he studied Spanish and law, but upon his return to the United States his interest turned toward astronomy, in which field he completed a thesis in 1917 based on observations at the Yerkes

Observatory. He served in the United States' army during the last phase of World War I and subsequently began work at the Mount Wilson Observatory, where he had at his disposal the 100-inch Hooker telescope. It was with this instrument that he made his breakthrough in international astronomy. In the autumn of 1923 he found an object in the Andromeda nebula which he identified as a Cepheid variable (he first thought it was a nova). Using the period–luminosity law he concluded that Andromeda was some 800,000 light-years away and thus a separate star system well outside the Milky Way. As a result of Hubble's discovery the "island universe" soon became generally accepted.

A specialist in what he called extragalactic astronomy, Hubble decided to investigate the correlation between the distances of the nebulae and their spectral redshifts. In a landmark paper of 1929 he reported a linear relationship between the two quantities, which eventually became known as the Hubble law. For the constant of proportionality (the Hubble constant) he found a value of about 500 km/s/Mpc. The following year, after Lemaître's theory had become known, the Hubble relation came to be seen as solid evidence for the expanding universe. A cautious empiricist, Hubble never unequivocally accepted that the universe is in a state of expansion. During the 1930s he started a programme in observational cosmology based on measurements of the number of galaxies and their brightness. The programme continued after World War II when the 200-inch Hale telescope became operational, but it was left to his assistant Allan Sandage to carry it through. During his distinguished career Hubble received nearly all the medals and honours a scientist can dream of. Only the Nobel Prize was missing.

Biographical sources: Christianson (1995); Mayall (1970); Smith (1990). For works on and by Hubble see <http://www.phys-astro.sonoma.edu/brucemedalists/hubble/HubbleRefs.html>.

1. *Grace is one of them*: In 1924 Edwin Hubble married Grace Leib, the widowed daughter of a Los Angeles banker. As Grace Burke Hubble she became of great importance to the astronomer, his career, fame, and social life.

2. *my paper of 1926 on the classification of the extragalactic nebulae*: At the end of this long and important paper, Hubble (1926) estimated the number of galaxies in a given volume of space to be 9×10^{-18} galaxies per cubic parsec, which he translated to a mean density $\rho = 1.5 \times 10^{-31}$ g/cm^3.

3. *I used formulae from a textbook*: Hubble relied on Arthur E. Haas' *Introduction to Theoretical Physics*, vol. 2 (London: Constable & Co., 1925), where the formula appears on p. 373. The book was a translation of a textbook Haas had published in German from 1919–1921. Based on Einstein's formula for a closed and static universe, Hubble found a radius of 27×10^9 parsec $\cong$ 10^{11} light-years and a total mass of 1.8×10^{57} g or 9×10^{22} solar masses. For Hubble's and other's attempts to find the size of the universe before 1930, see Peruzzi and Realdi (2011).

4. *conversations with Tolman and Robertson*: It is possible that Hubble knew of the theoretical predictions of a linear redshift–distance relation before or at about the time he wrote the 1929 paper. The information might have come from discussions with Tolman. However, he presented his work as a purely empirical investigation. Robertson, who in 1928 had published a linear redshift–distance relationship based on theory, stayed in Caltech as assistant professor between 1927 and 1929 and was in contact with Hubble. See Smith (1982, pp. 199–200) and Nussbaumer and Bieri (2009, p. 118).

5. *I actually met him in 1924 or thereabouts*: During his stay in the United States, Lemaître visited Vesto Slipher at the Lowell Observatory and Hubble at the Mount Wilson Observatory to learn about the most recent measurements of redshifts and about extragalactic astronomy in general.

6. *I did not actually conclude that the universe is expanding*: There is no indication in Hubble's 1929 paper that he interpreted the data to be the result of an expanding universe. He never claimed that this is what he thought. The paper was observational and the only theorist he referred to was de Sitter. See Kragh and Smith (2003).

7. *the de Sitter effect*: Hubble (1929a, p. 173) suggested that his data indicated the "possibility that the velocity–distance relation may represent the de Sitter effect." This effect, he explained, was a displacement of the spectra arising "from two sources, an apparent slowing down of atomic vibrations and a general tendency of particles to scatter." See also Chapter 5. He also thought that the origin of the redshifts might be gravitational, a possibility that he mentioned in a letter to Shapley of May 1929. See Kragh and Smith (2003, p. 151).

8. *apparent velocities*: Hubble presented what came to be known as the "Hubble law" as a relation between distance and radial velocity, as he phrased it in the title of his 1929 paper. However, he made it clear that the velocities v were merely apparent, obtained from the Doppler formula $\Delta\lambda/\lambda = v/c$, where $\Delta\lambda/\lambda$ is the redshift and c the speed of light. In a letter to de Sitter of 23 September 1931, he wrote: "We [Humason and I] use the term 'apparent' velocities in order to emphasise the empirical features of the correlation. The interpretation, we feel, should be left to you and the very few others who are competent to discuss the matter with authority." (quoted in Smith 1982, p. 192). Again, in a paper of 1931 co-authored by his assistant Milton Humason (Hubble and Humason 1931, p. 80) he wrote that it was "constrained to describe the 'apparent velocity-displacements' without venturing on the interpretation and its cosmologic significance." In another paper Hubble (1929b, p. 96) explicitly distanced himself from an interpretation in terms of a real Doppler effect. "It is difficult to believe that the velocities are real," he wrote.

9. *the work with Humason*: Together with Humason, Hubble published a much extended set of data, proving that the linear relationship was valid out to a distance of 32 Mpc and apparent radial velocity of 20,000 km/s (Hubble and Humason 1931). Whereas Hubble in 1929 estimated the constant of proportionality to be about 500 km/s/Mpc, in the 1931 paper it was found to be 558 km/s/Mpc.

10. *the velocity–distance relation is now called the "Hubble law"*: Until the early 1950s eponymous labels such as "Hubble's law" and the "Hubble constant" were rarely used. The first to raise the status of the velocity–distance relation to a law associated with Hubble's name may have been Arthur G. Walker (1933, p. 159). Bondi (1948) used the term "Hubble's constant," which in 1957 was so well established that it entered the *Encyclopaedia Britannica*. The same was the case with the symbol for the constant (H or H_0), which in the literature of the 1930s was often written as h or some other symbol. For example, h was the symbol used by the French astronomer Paul Couderc in a book of 1950 that was translated into English two years later. Couderc (1952) wrote of "Hubble's constant" and for the empirical velocity–distance relation he used both "Hubble's law" and the "Hubble–Humason law."

11. *a rose would smell as sweet*: "What's in a name? That which we call a rose / By any other name would smell as sweet." *Romeo and Juliet*, Act II, Scene 2.

12. *to designate it by the letter H*: Hubble (1929a) used the symbol K. He did not write down the Hubble relationship, but formulated it in words only: "The data in the table indicate a linear correlation between distances and velocities."

13. *without giving proper credit to my earlier formulation*: In a detailed paper of May 1930, de Sitter determined the constant in the velocity–distance relation to be 463 km/s/Mpc. Hubble complained that de Sitter had failed to recognize his earlier work of 1929 sufficiently, accusing him of having acted unethically. "I consider the velocity–distance relation, its formulation, testing and confirmation, as a Mount Wilson contribution and I am deeply concerned in its recognition as such," he wrote. On Hubble's "angry letter," see Nussbaumer and Bieri (2009, pp. 129–132) and Hetherington (1982, p. 48).

14. *not taken seriously by my colleagues in astronomy and physics*: The steady-state theory of the universe was either dismissed or ignored by American astronomers. According to Allan Sandage, who in the early 1950s worked as Hubble's assistant, a main reason was the steady-state theorists' reputed disdain of observations. He later said that when this became known "by the Mount Wilson astronomers, they just dismissed all the steady state boys. That was the beginning of the rejection." See Overbye (1991, p. 40) and Christianson (1995, p. 351). In an interview of 1978, Sandage said: "The difference in philosophy between Hoyle and Hubble was profound. Hoyle, I think, was and is more interested in the worlds that could be, instead of the world that

is, . . . whereas Hubble was an absolute empiricist, asking, 'What is it really like?' " (AIP interview 1978; see AIP Interviews).

Hubble seems to have referred only once to the steady-state theory. In a lecture of 1951, he called it "a bold hypothesis of continuous creation of matter." (Hubble 1951, p. 464). For the steady-state theory, see Chapter 11.

15. *recent observations by Stebbins and Whitford*: As quoted in Hubble (1951, p. 468). In 1948 the American astronomers Joel Stebbins and Albert Whitford found an excess reddening of galaxies increasing with their distance. The "Stebbins–Whitford effect" was generally seen as proof that the more distant galaxies were redder than those close by, thereby indicating an evolutionary difference between galaxies that could not be explained on the basis of the steady-state theory. See Kragh (1996, pp. 276–279) and also the interview with Gamow in Chapter 10.

16. *I don't remember having met him*: Hubble's memory failed him. He had first met Hoyle at a meeting of the International Astronomical Union in Zurich in August 1948, where Hubble presided over the Commission on Extragalactic Nebulae. Hoyle first visited Caltech in 1944 and again in the first months of 1953, when he came to know Hubble well. "On most Sundays while at Caltech, I had the pleasure of walking with Hubble and his wife Grace," he recalled. See Hoyle (1994, pp. 252 and 277).

17. *I even had some sympathy for Milne's universe*: Hubble had met Milne on several occasions in Oxford and elsewhere. He came to know him well during an extended visit that he and Grace made to Oxford in 1934, when Hubble was invited to give the Halley Lecture and Milne was one of his hosts (Weston Smith 2013, pp. 179–180). During the 1930s Hubble vacillated between a static and an expanding universe, both of which possibilities were included in Milne's theory, depending on which time-scale was used. In his Halley Lecture he paid tribute to "Professor Milne's fascinating kinematical theory of the expanding universe." (Hubble 1934, p. 17). Referring to Milne's *Relativity, Gravitation and World Structure*, he later wrote about the model that "it appears to possess unusually significant features." (Hubble 1937, p. 199).

For his part, Milne (1938, p. 344) believed that Hubble's observations provided empirical support for his theory: "Hubble's observations disclosed a density-distribution of nebulae increasing outwards if recession is adopted, and a homogeneous distribution if recession is denied. This is just what is predicted on the present treatment." Milne died in 1950 and his theory soon passed into oblivion.

18. *the dreamy realms of speculation*: The expression appears in Hubble (1936a, p. 202). In the introductory chapter of his book, Hubble stressed that theory needs to stay in close contact with observational data. "One of the few universal characteristics [of science] is a healthy skepticism toward

unverified speculations," he wrote. "These are regarded as topics for conversation until tests can be devised. Only then do they attain the dignity of subjects for investigation." (Hubble 1936a, p. 6).

19. *a lecture I gave earlier this year*: Hubble is referring to the R. A. F. Penrose, Jr, Memorial Lecture, which he delivered on 19 April 1951. See Hubble (1951, p. 470).

20. *cosmologists deal with possible worlds*: Hubble (1942b, pp. 104–105) spelled out his view as follows: "Mathematicians deal with possible worlds, with an infinite number of logically consistent systems. Observers explore the one particular world we inhabit. Between the two stands the theorist. He studies possible worlds but only those which are compatible with the information furnished by observers. In other words, theory attempts to segregate the minimum number of possible worlds which must include the actual world we inhabit. Then the observer, with new factual observation, attempts to reduce the list still further."

21. *Lemaître's model fares rather badly*: In a paper of 1936 Hubble critically compared observational data with "the type [of universe] that will always be associated with the name of Lemaître." According to him, "The high density suggests that the expanding models are a forced interpretation of the data." (Hubble 1936b, pp. 551 and 517). He also considered Lemaître's model in the Rhodes Memorial Lectures given in Oxford the same year, concluding that although it could not be ruled out, it was "rather dubious." (Hubble 1937, p. 62).

22. *values of a human or religious nature*: Hubble, who seems to have been uninterested in religion or was possibly an atheist, was careful to keep questions of a moral and religious nature outside his publications and lectures. For his view of science and human values see Hubble (1954). Contrary to Hubble, his friend Tolman (1934, p. 486) explicitly warned against philosophical preferences in cosmology. "We must," he said, "be specially careful to keep our judgments uninfected by the demands of theology and unswerved by human hopes and fears."

23. *cited by none other than the pope in Rome*: In a controversial address of 22 November 1951 to the Pontifical Academy of Sciences in Rome, Pope Pius XII argued that modern cosmology had provided convincing evidence for a beginning of the universe and, as a consequence, the existence of a divine creator. See Kragh (1996, pp. 256–259). Among the evidence cited by the pope was "The examination of various spiral nebulae, especially as carried out by Edwin W. Hubble [sic] at the Mount Wilson Observatory." The address quickly became widely known and discussed, for instance in *Time* of 3 December 1951, which carried an article titled "Behind Every Door: God" in which the pope's reference to Hubble was quoted. The Rome address can be read online at <http://papalencyclicals.net/pius12/p12exist.htm>.

24. *a letter an old friend of mine sent me*: A letter of 23 November 1951 from Elmer H. Davis, quoted in Christianson (1995, p. 348). Hubble received the text of the pope's address in a letter of 4 December.

25. *cosmology has almost become a Californian monopoly*: Sandage recalled: "Cosmology at that time was a one or two man subject, in the world. And those two or three people were Hubble, Humason, and Baade. Almost all the other astronomers every place else in the world were doing something else. . . . It was brought about in the 1950s this strong monopoly with other three or four people in the world being involved, by the accidents of the large telescope." (Sandage, AIP interview 1978; see AIP Interviews).

26. *It penetrates into space at least about twice as far as the 100-inch*: Hubble (1951) and Hubble (1953) give accounts of the Palomar instruments and their use in cosmology and extragalactic astronomy. The "100-inch" refers to the Mount Wilson reflector, whose main mirror had a diameter of 100 inches or 2.54 m.

27. *the new radio telescopes*: Radioastronomy as a research area dates from the early 1950s and was initially dominated by British and Australian astronomers. Hubble's scepticism about radio cosmology was understandable in 1951, but a few years later Martin Ryle and others recognized the relevance of radioastronomy to questions of cosmology, including its role in discriminating between rival cosmological theories. In the period from 1955 to 1965 radioastronomy had a greater impact on cosmology than the big optical telescopes in California. See Munns (2013) for a careful analysis of how radioastronomy and physics changed the astronomical scene in the 1950s.

28. *entrance of radio physics and engineering into astronomy*: Not all traditional astronomers viewed radioastronomy with the same distrust as Hubble. In a memorandum of 1952 Harlow Shapley, retiring director of Harvard College Observatory and Hubble's senior by five years, said: "The rich harvest of exciting knowledge that radio astronomy has yielded in recent years has been gathered almost exclusively by electronic physicists, men skilled in phototubes, circuits, and the intricacies of electronic science. The discoveries made by these wizards of the micro-waves are largely in the field of astronomy." (quoted in Munns 2013, p. 24).

29. *in charge of our two Californian observatories*: The Mount Wilson Observatory was owned and operated by the Carnegie Institution, while Mount Palomar was run by Caltech. In 1948 the two observatories merged administratively. When a new director of Mount Wilson had to be appointed in 1945, Hubble wanted the job, but he was passed over. He was also not considered for the directorship of Mount Palomar. Instead the physicist and astrophysicist Ira Bowen was named director, much to Hubble's dissatisfaction. Wanting to keep physicists out of the large observatories, in 1945 Hubble protested against "the appointment of a physicist as director of the astronomical

center of the world." See Osterbrock (1992) and also Christianson (1995), pp. 304–310.

30. *Hale started the project*: The American astronomer George Hale did important work in solar astronomy and was a key figure in the foundation of several astronomical observatories, including the Mount Wilson Observatory and the Palomar Observatory. In 1928 he secured money from the Rockefeller Foundation for the construction of what twenty years later would become the Palomar Observatory.

31. *if there is a deceleration*: The so-called deceleration parameter q_0 was introduced in the mid-1950s, after Hubble's death. It is a dimensionless measure of the rate of slowing down of the expansion that can be determined by plotting the redshifts of galaxies against their apparent magnitudes. For models with $\Lambda = 0$, $q_0 = \frac{1}{2}$ implies a flat space as in the Einstein–de Sitter model; if $q_0 > \frac{1}{2}$ space will be closed and if $q_0 < \frac{1}{2}$ it will be open. For a careful and critical examination of Hubble's research programme and a comparison with that in the post-Hubble era, see Sandage (1998).

32. *I have talked with Sandage about it*: Together with Humason and Nicholas Mayall, Sandage used the redshift–magnitude method to determine $q_0 = 2.5 \pm 1$, indicating a universe different from the one prescribed by the steady-state theory, where $q_0 = -1$. (see Humason et al. 1956). Allan Sandage served at the time as Hubble's assistant, and after 1953 he succeeded him as head of the Mount Palomar research team. See also Chapter 11.

33. *the expanding universe was not vindicated by available data*: Hubble's public announcements were agnostic. "Because the telescopic resources are not yet exhausted," he said in *The Realm of the Nebulae*, "judgement may be suspended until it is known from observations whether or not red-shifts do actually represent motion." (Hubble 1936a, p. 122). Several times Hubble expressed scepticism with regard to the expanding universe. For example: "The empirical evidence now available does not favor the interpretation of red shifts as velocity shifts," and "The interpretation of red-shifts as velocity-shifts is less satisfactory." (quotations from Hubble 1942a, p. 214 and Hubble 1951, p. 463, respectively). According to Sandage (1998), Hubble maintained his scepticism with regard to the reality of the expansion until the end of his life. On the other hand, Norriss Hetherington (1982) has argued that philosophical principles caused Hubble to prefer an expanding and homogeneous universe of the type proposed by Lemaître, despite such a model not fitting well with his data.

34. *the extensive work that Tolman and I did*: A reference to Hubble and Tolman (1935), who concluded (p. 335) that observations could be explained "on the basis of either a static homogeneous model with some unknown cause for the red-shift or an expanding homogeneous model with the introduction of effects from spatial curvature which seem unexpectedly large but may not

be impossible." The Hubble–Tolman paper formed much of the basis of Hubble (1936a) and Hubble (1937).

35. *Zwicky and some other physicists*: The Swiss-American astrophysicist Fritz Zwicky was professor at Caltech and a participant in astronomical discussions with Hubble, Tolman, and other scientists. As early as 1929 he suggested that Hubble's redshift–distance relation could be explained by the hypothesis that photons lose energy during their journey through space. The class of "tired light" hypotheses continued to be defended by a minority of astronomers for the next several decades.

36. *some hitherto unknown principle of nature*: See, for example, Hubble (1942a, p. 214) and Hubble (1947, p. 165). He did not elaborate on the nature of the unknown principle of nature except that he associated it with an infinite and homogeneous universe.

37. *it's problematic for methodological reasons*: Hubble (1937, p. 26) expressed his reservation as follows: "We cannot assume that our knowledge of physical principles is yet complete; nevertheless, we should not replace a known, familiar principle, by an ad hoc explanation unless we are forced to that step by actual observations."

10

George Gamow: Nuclear Physics and the Early Universe

Interview conducted on 22 October 1956 in Gamow's office at the physics department, University of Colorado, Denver. Language: English.

Starting in 1939 with the irresistible *Mr Tompkins in Wonderland*, I had read with great pleasure the popular books by George Gamow. By the early 1950s I was well aware of his arguments for a universe born a finite time ago in an inferno of interacting nuclear particles. Although sympathetic to his idea and what I considered its sound basis in nuclear physics, I also understood that it suffered from various weaknesses and was far from enjoying general recognition. When I asked astronomers about it, they typically shrugged their shoulders, dismissing it as speculative and irrelevant. There were even those who had never heard of Gamow's cosmic explosion hypothesis.

When I first contacted Gamow he was still at George Washington University in Washington, DC, where he had been for about two decades. But he was planning to move to Denver in beautiful Colorado, and that is where I met him in the autumn of 1956. As he told me, he had not yet accommodated to the new setting and generally felt disturbed by the changes in his professional and personal life (which included a recent divorce from his wife, to whom he had been married since 1931). Nonetheless, he was relaxed and forthcoming, and often joking, during the interview, apparently quite happy to tell me about himself and his important work in cosmology. I instantly liked the big man (Figure 11). Contrary to other scientists I had engaged in conversation, he did not wait for me to start the interview. And whereas Hubble had offered me lemonade, Gamow insisted on bourbon.

On my way back from Denver I learned that the Suez crisis had escalated to warfare, with Israeli forces attacking Egyptian positions in the Sinai Peninsula. In spite of the disturbing news, I was thinking about the universe and what Gamow had told me about its birth and fate in the far future.

Figure 11 George Gamow in the late 1930s. The writing on the photo is in Danish: "With many heartfelt greetings from Geo. Gamow." Courtesy of the Niels Bohr Archive, Copenhagen.

GG Welcome to Denver, Mr Nielsen, I understand you are from Denmark. I have fond memories of Copenhagen[1] from the time I stayed at Bohr's wonderful institute, where there were more than a few Nielsens, if I recall correctly. I actually learned to speak Danish and still remember bits of it. Should we try to speak in Danish? That might be fun.

CCN Fun perhaps, but let's do it properly and keep to English. I understand why you liked Copenhagen, it's a nice city. Now, Professor Gamow: you are a Russian-American nuclear physicist and also a pioneer in cosmology, and to the public you are known and much appreciated as a writer of popular science books. Nuclear physics deals with the inconceivably small atomic nucleus and cosmology with the inconceivably big universe, so perhaps you could start telling me how you came from one end of the distance scale to the other? It seems to be as big a quantum jump as one can imagine.

GG Only on the face of it. Nuclear physics and cosmology are in fact intimately connected, and especially as far as the very early universe is concerned—it's really nature's own nuclear laboratory, or rather it was a couple of billion years ago. I'm a physicist with many interests and for the last decade cosmology is a topic that has loomed large, but I'm also interested in the new science of molecular biology[2] and have recently come up with some ideas that you . . .

CCN Yes, I'm sure you have, but for now could we please focus on your contributions to cosmology?

GG Sure, perhaps we can return to biology, it's so interesting. But wait, you'll be interested to know that just recently a friend of mine showed me a passage from a letter that Charles Darwin wrote to a colleague of his, a zoologist perhaps, some time after he had completed his great work. And he says something like, "one cannot think of the origin of life, it's as incomprehensible as the origin of matter."[3] Yes, I think that's pretty close to what Darwin said, and it's funny, isn't it, for that's precisely what I am doing, thinking *both* of the origin of matter and the origin of life. Darwin would have been surprised if he returned from his grave, wouldn't you say?

CCN I bet he would, but . . .

GG Great! I knew you would appreciate my point. As I was saying, my own route to cosmology was via astrophysics and my wish to understand the origin of the elements of which all matter is made. How is it that the cosmic distribution of the elements follows a distinct pattern[4] and that there is so much helium in the universe and so little of the heavy elements? I began thinking of this in the mid-1930s, and in 1940 when I wrote my book on the Sun[5] various ideas began to crystallize in my mind, if only slowly. My way of reasoning was much like that of archaeologists when they draw conclusions about past cultures from the remnants left from them. I consider myself a nuclear archaeologist, for I try to reconstruct the cosmic past on the basis of what is left over from it in the form of atomic nuclei and perhaps some radiation of low intensity.

CCN A nuclear archaeologist! That's some name.

GG Yes, isn't it. Well, at the time I didn't connect these ideas with Friedmann's equations, at least not clearly, but they must have been floating around somewhere in the back of my mind, for I knew the equations very well from my training in Russia. You see, I was a student of Friedmann[6] and followed his lectures in St Petersburg—now, of course, it's called Leningrad, but I don't like that name at all and prefer calling it St Petersburg. Where was I? Yes, Friedmann had this idea of an expanding universe that started in a state in which the density and temperature of matter were huge[7]—nearly infinite. But it took me some time before I connected Friedmann's brilliant insight with my own ideas on the formation of elements.

It was during the war, you know, and I arranged a conference in Washington[8] where we agreed that the nuclear processes resulting in the elements could not be of the equilibrium type but had to be an explosive event at the beginning of time. Well, perhaps not all of us agreed, but I agreed! And a few years later I wrote up a paper in which I proposed that initially the entire universe had been a compact soup of neutrons[9] that began expanding in agreement with Friedmann's equations. That was shortly after the war, in 1946 I think, and that was the beginning of it. Oh, would you care for a drink?[10]

CCN No, thank you, it's too early. Tell me, this is the same picture of the origin of the universe that you outlined in your popular book on atomic energy,[11] isn't it? I read it with great pleasure.

GG Yes it is, indeed it is. But of course it was for the consumption of a lay public and not aimed at my colleagues in physics.

CCN Now, I notice that you refer to the Friedmann equations rather than the Friedmann–Lemaître equations, and that leads me to ask about the connection between your ideas on the origin of the universe and those of Lemaître. After all, aren't they basically the same? Could one say that your theory of the explosive universe is a more refined version of the primeval atom hypothesis?

GG There actually is no connection. Although I knew of his theory it didn't impress me. The primeval atom is just another name for Friedmann's early universe and it cannot be turned into a quantitative model based on nuclear reactions. No, my ideas were quite independent of Lemaître's.[12] It makes no sense to speak of a Lemaître–Gamow theory, as someone suggested to me. At any rate, I have no contact with him and he seems no longer to be active in cosmological physics.

CCN I interviewed him back in 1938, at a conference in Notre Dame. Have you ever met him?

GG No, I haven't. We belong to two different generations and two different scientific cultures.

CCN So, your paper of 1946 was the beginning of your theory of the creation of the universe that you put forward in your famous book[13] about five years ago?

GG Yes, it was, and I think Einstein liked it.[14] I wrote him a letter in which I described my idea and some other ideas I had about the universe—some of them correct, others not. I don't mind if nine out of ten of my ideas are wrong,[15] as long as the tenth is right and fruitful. A success rate of ten per cent isn't so bad, wouldn't you agree?

But the project only gained momentum when I suggested to Ralph Alpher that he should write his doctoral dissertation on the formation of elements in the primeval universe. His thesis attracted a lot of attention,[16] you know, and about that time he started collaborating with me on a joint paper in which we modified and developed the original scenario of the early universe into a more realistic version. We were actually able to obtain a fairly good agreement with Goldschmidt's abundance data, which was a promising start. This paper of ours is often referred to as the $\alpha\beta\gamma$ paper.[17] Do you know why?

CCN No, please enlighten me.

GG Oh, it was really a joke,[18] and a very good one I think. You see, Alpher is just like alpha (α) and Gamow is like gamma (γ), so it occurred to me that beta was missing from the start of the alphabet. And what did I do? Well, I submitted our paper to *Physical Review* adding Bethe as an author—in absentia of course! That turned it into the $\alpha\beta\gamma$ paper, and most people had no idea it was a pun. I even succeeded in smuggling in a fake reference to "Delter" in the issue of *Reviews of Modern Physics* celebrating Einstein's seventieth birthday, thus making it an $\alpha\beta\gamma\delta$ theory. And just recently I invented Mr Tompkins as my co-author in a paper on the genetic code, but unfortunately it didn't work this time. I could tell you of other stories . . .

CCN Yes, I'm sure you could. It's all very amusing, but perhaps we should keep to the content of the papers. I understand that . . .

GG But, oh, I must tell you about one more joke, one you have probably never heard, but you may know that some years ago the pope in Rome gave an official address in which he used my theory to prove the Christian creation story. Well, he didn't actually refer to me, but he did use the theory I had worked out with Alpher and Herman. He can do as he pleases and I know that the guys in the Vatican are interested in it.[19] Anyway, I decided to introduce a technical paper with a lengthy quotation from the pope's address,[20] just for fun, as if he were a recognized physicist. It's probably the first time His Holiness has been cited in *Physical Review*!

CCN That's great, I know of the pope's intervention in cosmology and also that Hubble got involved in it. He told me about it when I interviewed him. But now I would like to ask you a few questions about how you conceive of the universe, I mean apart from its origin in a nuclear explosion. As far as I understand you have no confidence in the cosmological constant and you believe, contrary to what used to be the standard view and perhaps still is the standard view, that the universe is open and infinite—have I got it right?

GG Well, yes, I never liked the cosmological constant and I much agree with Einstein who once told me that it was the biggest blunder in his life as a physicist.[21] The only excuse for keeping the constant is that it can produce a longer lifetime for the universe and thus avoid the intolerable paradox of stars that are older than the universe. But this excuse is no longer valid after Baade showed that the Hubble constant is much smaller[22] than we thought. Incidentally, this also removes the main argument in favour of the steady-state theory, although Hoyle won't admit it.

An infinite universe, you said? Yes, it seems to me that observations indicate that the curvature of space is negative—we live in a hyperbolic universe,[23] on the surface of a saddle as it were. Not that I'm married to the infinite universe, it just seems to be what nature has chosen. No reason to be afraid of infinity, if you ask me. People tend to find it absurd that matter in an infinite space can contract or expand and still occupy the same infinite space, but as Hilbert showed a long time ago,[24] that's how infinity is. Are you sure you don't want a drink?

CCN Yes, I'm quite sure, thank you. I must admit I find it hard to conceive of an infinite universe, it sort of makes me dizzy . . . , and there are all these conceptual problems with an infinite number of stars that makes it hard to believe that it's real.

GG Yes, when I'm in my serious mood—and it happens from time to time—I reckon that it may be impossible to decide whether or not the universe is truly infinite. I remember that Stan Ulam once told me about Gödel's theorem[25] that certain mathematical statements are unprovable. It's possible that the question is one of those statements that in principle escapes decision. But we have to make one assumption or other, and I find the infinite universe to be more elegant than the finite one. We all have our preferences and that's mine.

CCN Elegant? Hmm, I see. But tell me, do we just have to accept the beginning of the universe as a postulate that cannot be explained on the basis of science, or is it possible that there is a reason why it started to exist some billion years ago? Could there be something before the beginning that explained it? I know it sounds silly . . .

GG It's not silly at all, Mr Nielsen. In fact that's the way I usually think about the universe, and I'm not the first to do so. In the early days of Christianity there were pious people wondering what God was doing before he made the world,[26] and St Augustine told them, "He was making hell for those who ask such foolish questions!" Well, it may not be a literal quote, but he said something like it. My point is that although the question may be heretical, it's not foolish. It's quite possible, and I would even say likely, that our expanding universe is the result of an earlier contraction[27]—that the universe bounced instead of being created, in which case it is infinitely old rather than just a few billion years old.

CCN Aha, so in the far future there will be another bounce, and so on?

GG No, there won't. That's what people call the cyclic universe and it doesn't agree with observation. For about five billion years ago the universe contracted to an incredibly dense state, or perhaps it did—it's just a hypothesis—and then it began expanding and it will continue to expand in an eternity of time. I coined the term "big squeeze" for the state of maximum compression, but there will never be another big squeeze, at least we have no reason to believe there will.

You probably know that it all depends on the density of matter in the universe. There is the remote possibility that there are large amounts of dark matter in intergalactic space[28] and if so the cyclic universe will be a possibility. We can't tell for sure, not even if there was a universe before ours, for all information from this earlier state would be erased in the big squeeze. All this is more metaphysics than physics, I admit that. But then, I don't mind crossing the border to metaphysics from time to time, just for the fun of it.

CCN The big squeeze, yes, I remember the name, it's sort of similar to the "big bang" that Hoyle invented as a not very friendly label for the kind of explosive universe you and your colleagues favour. It hasn't really caught on, I think, but when people use it, they have your theory in mind.

GG I'm aware of the name, of course, but I don't like it at all, it's a phrase that gives quite a wrong impression of the primordial universe and its early expansion. I have never used the name myself[29] and I don't intend to use it in the future.

CCN Whether we call it a big bang, a big squeeze, or something else, in the early inferno we have nuclear particles interacting at very high temperature, producing simple atomic nuclei and eventually not-so-simple nuclei. I think you estimate the temperature shortly after the explosion to several billion degrees—much greater than the temperature in the centre of our Sun, if I'm right.

GG You're right. It seems that one billion absolute degrees is a kind of fundamental unit of temperature.[30] It really ought to have its own name—do you have a suggestion, Mr Nielsen?

CCN A temperature unit for the initial inferno? Well, . . . why not call one billion degrees "one inferno"? I'm not serious, of course, but since you asked me . . .

GG One inferno? Hmm, that's not a bad idea, and Dante would have been pleased, I think. I'll consider it.

CCN In this initial state, inferno or whatever, I understand there's also a lot of radiation?

GG Oh yes, in the very early and very hot phase there is much more energy in the form of radiation than in the form of matter, very much more. That's because the radiation energy density increases steeply with the temperature. If the temperature doubles, the radiation density will increase by a factor of sixteen. But as the universe expands it also cools and after some time the matter density will exceed the radiation density[31]—Today, of course, we live in a cold and matter-dominated universe, something we should be grateful for. You wouldn't like to live in a radiation-dominated universe, I can tell you that. Cheers!

CCN What has happened to the radiation? Are there still relics of the primordial radiation, only more rarefied and colder than in the past? Isn't that what you say in your book?

GG Yes, it's in my book. It's almost like in the Bible,[32] you know, with its claim that "in the beginning there was light," except that in the real universe the radiation was in the invisible high-frequency range, X-rays and that kind of stuff. As to the present radiation, it was something we calculated in the late 1940s, perhaps it was Alpher and Herman at first, but we did most of it collectively.[33] I've just published a paper in which I get a temperature of about 6 K,[34] which is of the same order as the heat generated by the stars, but I don't think the two kinds of radiation can be distinguished experimentally. It's possible that Alpher and Herman think differently,[35] though. In any case, our colleagues in astronomy and physics don't seem to care, they have just ignored our prediction, I don't know why. Perhaps it's not so important, for there are other and better reasons showing why our theory is essentially correct.

CCN Would that be the calculation of primordial helium, for example?

GG Yes, when I started out in this business it was to account for the abundance of the elements, and we have succeeded in calculating that in less than half an hour after the beginning of time about one-quarter to one-third of all nucleonic matter transmuted into helium isotopes. We also get a small amount of deuterium, but unfortunately we have no clue about how much deuterium there is in the universe.

A Japanese by the name Hayashi[36] also did some work in the area, but it was on the basis of our theory. He improved it, sort of. The figure of one-quarter is in fair agreement with the observational estimate, although at present we don't know the amount of helium precisely.[37] All the same, it's a major success of our theory that we can account for ninety-nine per cent of all matter[38] in the universe. Not bad, wouldn't you agree?

CCN But critics point out that . . .

GG Yeah, yeah, that we cannot synthesize the heavier elements, lithium and beyond, and that's true enough, it's the famous mass gap problem.[39] Oh, we worked hard on it, but whatever we did it didn't work, so it's possible that the mass gap cannot be bridged under the circumstances governing the early universe. But does it mean that the theory must be wrong? No it doesn't, for it's possible that there is a way to bridge the gap, a way that we just haven't found yet. More to the point, perhaps, even if we are forced to admit that the heavier elements are formed in the stars and not cooked in the primordial nuclear oven, so what? It wouldn't ruin our theory, for it is consistent with a mixed origin of the composition of matter,[40] some of it being of stellar origin and some of cosmological origin. People don't seem to realize that, they speak of the heavier elements as if they contradict the theory. That's plain wrong. They don't.

CCN Yes, that's a good point. Now, among the critics of your theory are the steady-state people, of course, and I can imagine that you are no less critical of their ideas than they are of yours. What, in your view, are the main weaknesses of the steady-state theory?

GG I'm surprised that it's taken seriously at all, but then it's mostly in England, I think. Britons have their own ideas, they are funny people. Well, on the observational side, just to mention two examples, it cannot explain the amount of helium and it also cannot explain the formation of galaxies and their different ages.[41] If one follows the logic of the theory our own Milky Way must be extraordinarily old and we would have to go out *very* far in space until we met another galaxy of the same age. The theory really can't explain the evolutionary features of the universe as a whole, and the reddening effect found by Stebbins and Whitford[42] directly contradicts it.

CCN As far as I know, . . .

GG And that apart, on the theoretical side the theory postulates a universe that doesn't change in time, it doesn't explain it on the basis of physics but merely supports the postulate by another postulate, the continual creation of matter. I could go on. But there's no point engaging in a controversy with Hoyle and his colleagues, the question will be decided by observations—in my view it has already been decided.

Incidentally, and this is not meant as support of the theory, many people are upset that the creation of matter in the steady-state theory violates the principle of energy conservation, as if it were a serious crime. This is not my view, for it's conceivable that this principle is wrong or only approximately valid. In fact, in my youth I was for a period quite enthusiastic about Bohr's suggestion of energy non-conservation,[43] so this part of the steady-state theory fails to shock me. Even Dirac suggested in the 1930s that energy is not strictly conserved, did you know that?

CCN No, I didn't. Now, there are rumours that you participated in a BBC radio debate with Hoyle,[44] but I have been unable to confirm it.

GG No wonder, for I didn't. Hoyle made his much publicized radio broadcast in 1950, but I had nothing to do with it. It may interest you that I had several discussions with Hoyle this summer in California[45] about the synthesis of elements, the background radiation, and other matters. Hoyle was staying at Caltech and I had a consultancy job for General Dynamics in La Jolla. I like him a lot, but I don't agree with him. Might it be because he's a Briton? I'm just kidding.

CCN I would like to know something about your general attitude to cosmology as a physical science. The question has been much discussed in England in particular, and there seems to be some disagreement as to the scientific status of cosmology, indeed about whether the study of the universe can ever be a science on par with other branches of physics—or astronomy, for that matter.

GG Okay, I'll be serious now. It's true that cosmology is very different from, for example, nuclear physics. There is no consensus on foundational questions in cosmology, witness the recent debate between evolution theory and the steady-state theory, but at least we all agree that in the end observation has priority over theory . . . well, now I think of it, I'm not so sure that the advocates of the steady-state theory agree. Well, to my mind there's no doubt that cosmology *is* a physical science and that it has no need of either philosophy or new physics. Personally I have no patience with those theories that are not founded on Einstein's equations[46] with or without the cosmological constant. Some would call me conservative, but the universe can surely be understood in terms of the known laws of physics such as relativity theory, quantum mechanics, and thermodynamics. I mean, if it can't, then it cannot be understood at all.

I consider myself a factual cosmologist;[47] you may even say that my attitude is that of an engineer's—engineers don't question the known properties of matter or the laws that describe it, they just use them, and so do I. It's a pragmatic attitude that is very different from that of the postulatory cosmologists in England.[48]

CCN But can one really avoid questions of a philosophical nature in cosmology? There is the question of the infinitely large universe, for example, and there is the question of a creation or beginning of it all. Already Kant[49] . . .

GG Spare me! Forget about Kant, we live in the twentieth century. No, it's a matter of physics, not philosophy, and should problems turn

rief6

up that apparently have no scientific answer then one would be wise to circumvent them for the time being. As I said, there is no absolute creation[50] in the kind of cosmology I suggest, so I don't have to speculate on something out of this world that caused the universe to come into existence. As to infinity, what has it to do with philosophy? It's simply a question of comparing theory and observation. No, just as we have long agreed to keep theology out of cosmology, so we should agree to avoid school philosophy.

CCN That is most illuminating, thank you. Now the explosion theory proposed by you and your collaborators has been around for nearly ten years and it might be interesting, just to round off our conversation, to look at the impact it has had. Are you satisfied with its reception among astronomers and physicists?

GG No, I can't say I am. Intellectually it has been most rewarding, but sociologically, so to speak, it has been a disappointment, I would even say a failure.[51] It attracted a good deal of attention in the period from about 1948 to 1952, but then interest in our theory diminished to nearly zero—within a year or two it came to a halt, and that's where we are now. Even in the days when people like Fermi and Wigner[52] took an interest in it, interest was limited to nuclear physicists. I think it was ignored in Europe and it was also pretty much ignored by astronomers, on both sides of the Atlantic. It is as if astrophysicists shy away from non-equilibrium nuclear processes and astronomers don't recognize the beginning of the universe as part of what they accept as cosmology. I'm not aware of a single paper in the *Astrophysical Journal* dealing with our theory of the explosive universe, but then . . . perhaps Chandrasekhar considers it too speculative.[53] I wouldn't be surprised.

CCN Have you actively promoted the theory?

GG Oh yes, it wasn't for lack of trying on our part. We really did what we could to spread the gospel, but apparently we failed. There was a conference in Belgium,[54] it must have been three years ago, and Alpher told me about it. He and Herman were there, speaking about our theory, but it was either resisted or ignored. Response was cool. At about the same time there was a summer school in Ann Arbor[55] to which I was invited. I gave lectures on the expanding universe, the origin of the chemical elements, the formation of galaxies, and some other subjects. It was a fine symposium and people enjoyed my lectures, but I don't think I convinced anyone that the universe started in a thermonuclear inferno. Since then nothing much has happened.

CCN Why do you think that interest in the theory has waned? Is it because of the competition from the steady-state alternative?

GG It's one factor, but not one of great importance, I think, after all it only enjoys very limited support. More important may be the widespread propaganda that our failure to synthesize the heavier elements amounts to a disproof of the theory, but as I told you, that is a mistake. The theory is alive, it's just in a phase of hibernation. I don't know, perhaps the reason for its decline is rather to be found in personal and social factors, in the way physics and astronomy develops nowadays. I can't really formulate what I mean . . .

Alpher and Herman have gone into industry[56] and have little time to do cosmology. I'm still in contact with them, but we don't collaborate any longer and I guess that some of the blame is mine. I mean, I haven't seriously tried to develop the theory. For the last few years I have been more interested in other things, not least in my work as an amateur biologist, so I'm a bit on the side-lines[57] when it comes to cosmology, but then, who isn't? I have also not been successful in recruiting young people to invigorate and promote the theory of the exploding universe, or rather, I haven't even tried to. It is as if it doesn't appeal to either physicists or astronomers—and who else should it appeal to? The theologians? I don't know and for the moment I don't really care. It's just like that, *c'est la vie* as they say in France.

CCN The theory is widely associated with your name and reputation as a physicist, and I wonder if that has been an advantage or disadvantage. For example, you are a prolific writer of popular science and many of your more recent contributions to the cosmological literature have been in the popular format. Although you may not have succeeded in convincing the majority of scientists, you have been quite successful in spreading the gospel to the general public, it seems. Could you say a few words about the relationship between popular and scientific literature, as you see it?

GG No need to go into detail, I think. Popular science is an old love of mine and, as you say, it has been much appreciated, perhaps more than my scientific work. In fact, I just recently received something called the Kalinga Prize.[58] It's a UNESCO prize for popular science writing—I had no idea there existed such a prize, but the money was most welcome and I didn't mind the honour either.

Well, unlike most of my colleagues I have no respect for the strict border that is supposed to distinguish research writings from popular writings. On the contrary, I often use popular writings[59] to present things in the simple way in which I think they should also be presented to other scientists. That's what I did in *The Creation of the Universe*, which I consider a hybrid between the popular and the scientific. This is unusual, I know, but I'm just continuing in my own way a style of science writing that Eddington cultivated with great success before the war. By the way, I met him at a meeting in Poland in about 1939; he gave a talk that I didn't understand a word of.[60] I liked the conference but was a little uneasy about being so close to the Russian border.

CCN Yes, I once talked with Eddington about that Polish conference and also about his work in popular science.

GG To return to your question, if you are suggesting that this atti-tude of mine may have harmed my academic reputation,[61] and by implication may have caused physicists and astronomers to consider the explosion theory of the universe as something they don't have to take seriously, well, there may be more than a grain of truth in it. But I can't change my style of doing science, and even if I could I wouldn't. Academic respectability, my ass! By the way, in this respect Hoyle is not so different from me, he also writes popular science and is currently working on a science fiction novel,[62] so he told me. We both have this feeling of being regarded as something of an outsider to the holy order of science just because we also write for the general public. It's absurd!

CCN One last comment and question, if I may. Cosmology is perceived somewhat differently in the Western and Eastern parts of the world, and some would say that it indicates its immaturity as a science. As a former citizen of the Soviet Union, do you know how your ideas of the early universe have been received in that country?

GG Oh, I prefer not to think about it, it's a misery. You know, Stalin has fortunately died, but even today science is heavily politicized. And I'm speaking not only of genetics but also of the cosmological mod-els that are deemed unacceptable for political reasons[63]—mine is one of them, and I'm proud that it is. As for myself I'm a non-person in my former fatherland—I don't exist and neither does my view of the universe. My books and articles are not translated into Russian and if I'm mentioned at all it's as a warning of Western scientific decadence.[64] Any kind of cosmological model with a finite past in censured by the ideological commissars, and scientists . . . well, the whole thing is a bizarre mixture of parody and tragedy, but that's what dialectical materialism is like.

I'm not really interested in politics, but I'm grateful to live in a free country that takes a firm stand against communism.[65] Not a word more about politics!

CCN No, not a word more, about politics or anything else! Thanks, I think this makes a good end to a conversation from which I have learned an awful lot. By the way, Professor Gamow, I wouldn't mind that drink now.

GG So now you come to your senses, Mr Nielsen! Cheers!

Notes

George Antonovich Gamow was born in Odessa, Ukraine, on 4 March 1904 and died in Boulder, Colorado, United States, on 20 August 1968. He began his education in Odessa and then moved to Petrograd (St Petersburg) to study physics, where he became part of a circle of young physicists including Lev Landau, Vladimir Fock, and Matvei Bronstein. Apart from learning the new quantum mechanics, he also attended lectures by Alexander Friedmann on relativity theory and cosmology. After graduation from what had become the University of Leningrad, in 1928–31 he went on a fellowship to western Europe. In Göttingen he made his first major contribution to physics, explaining in 1928 how alpha radioactivity could be understood in terms of quantum mechanics. He also went to Cambridge, England, and to Copenhagen, where he became an important member of Bohr's group. His expertise in nuclear physics resulted in 1931 in the pioneering textbook *Constitution of Atomic Nuclei and Radioactivity*. After his return to the Soviet Union he was faced with difficulties due to Stalin's repressive regime, and in 1933 he decided to leave the country. The following year he arrived in the United States to take up a position at George Washington University.

Gamow's research in the 1930s focused on astrophysics, including the problem of how to explain the relative abundances of elements by means of nuclear processes. In the summer of 1935 he proposed that neutron capture might be the basic mechanism in the formation of heavier elements in the stars, but he later concluded that the elements were of cosmological origin. His suggestion of 1946 that the early hot and dense universe was the site for the building up of elements was arrived at independently of Lemaître's earlier idea. In the years 1946–52 he developed, together with his assistants Ralph Alpher and Robert Herman, the theory that later became known as the big bang theory of the universe. Their theory accounted for the formation of helium and predicted a fossil microwave background, but could not explain the heavier elements. Although Gamow continued to work in cosmology, after the mid-1950s he increasingly turned to molecular biology. In 1956 he moved from George Washington University to the University of Colorado, Denver, where he stayed until his death (see Figure 11). Apart from his scientific work, Gamow was an outstanding popularizer of science, his books in this genre including classics such as *Mr Tompkins in Wonderland* (1939) and *The Creation of the Universe* (1952). In 1956 he received UNESCO's Kalinga Prize for his important contributions to the popularization of science.

Biographical sources: Gamow (1970); Reines (1972); Harper (2001); Chernin (1994). A list of Gamow's publications can be found in Harper et al. (1997, pp. 141–151).

1. *fond memories of Copenhagen*: From September 1928 to May 1930, and again from September 1930 to May 1931, Gamow stayed at Bohr's institute. He was a colourful and treasured member of the Copenhagen physics community, both scientifically and socially.

2. *interested in the new science of molecular biology*: Following James Watson and Francis Crick's famous discovery in 1953 of the double-helix structure of DNA, Gamow became greatly interested in the problem of finding the sequence of amino acids that would solve the problem of the genetic code. He published a couple of papers on the subject, which he took very seriously. In 1956, at the time of the interview, he considered his "extravagant deviation into the field of biological sciences" more interesting than nuclear physics and cosmology. See Gamow (1970, pp. 144–148) and Watson (2002).

3. *it's as incomprehensible as the origin of matter*: Quoting from memory, Gamow's reference was inaccurate if correct in substance. In a letter of 29 March 1863 to his close friend, the botanist Joseph D. Hooker, Darwin wrote that "It is mere rubbish thinking, at present, of the origin of life; one might as well think of origin of matter." See the letter at: <http://www.darwinproject.ac.uk/entry-4065>.

4. *the cosmic distribution of the elements follows a distinct pattern*: The relative abundance of nuclear species in the universe was approximately known in the late 1930s, after the Norwegian geochemist Victor Goldschmidt published an extensive survey on the subject. Goldschmidt's data were of great

importance during the early development of big bang cosmology (Kragh 1996, pp. 97–141). They showed that until the middle of the periodic system the abundance decreased rapidly with the atomic weight, after which it stayed at a roughly constant level.

5. *my book on the Sun*: In *The Birth and Death of the Sun*, a popular book completed in late 1939, Gamow found it supported by "good physical reality" that the early universe started in an extremely dense and hot state; as a result of the expansion, the density and temperature would decrease. As evidence for the scenario he referred, like Lemaître had done almost a decade earlier, to the order-of-magnitude equality of the Hubble time and the lifetimes of thorium and uranium. He spoke of "the creation of the universe from a primordial superdense gas." (Gamow 1940, p. 201).

6. *I was a student of Friedmann*: While a student at the University of St Petersburg, in 1923–4 Gamow followed a course by Friedmann on the mathematical foundation of the theory of relativity. He planned to write his diploma work on relativistic cosmology, but Friedmann's premature death in 1925 put a stop to it. Although Friedmann was thus not Gamow's supervisor, Gamow later referred to him as his teacher. See Frenkel (1994). From 1914 to 1924 the official name of St Petersburg was Petrograd, and from 1924 to 1991 it was called Leningrad.

7. *a state at which the density and temperature of matter were huge*: This is what Gamow (1970, p. 141) said in his autobiography. In fact, Friedmann's theory of 1922 had nothing to say about the density and temperature of the early universe. Gamow apparently read his own theory into that of his former teacher.

8. *I arranged a conference in Washington*: The Washington Conferences on Theoretical Physics, taking place 1935–42 and 1946–7, were jointly sponsored by George Washington University and the Carnegie Institution of Washington. For a list of the conferences and their participants see Harper (2001) and Harper et al. (1997, pp. 12–15). The eighth Washington Conference was devoted to "The Problems of Stellar Evolution and Cosmology." It was held from 23–25 April 1942 at George Washington University in Washington, DC, where Gamow was professor. The conference report stated that, to explain the abundance of heavier elements, "one must necessarily assume that two or three billion years ago the density of matter in space exceeded ten million times that of water and the temperature was as great as several billion degrees." It added: "These are just the conditions corresponding to the early stages of the expanding universe." See Gamow and Fleming (1942, p. 580).

9. *the entire universe had been a compact soup of neutrons*: Gamow (1946a) imagined the early universe as a dense mass of rapidly expanding neutrons which would coagulate into larger beta-radioactive complexes. As a result of neutron decay, they would turn into the atomic nuclei of the chemical elements.

He emphasized that "the conditions necessary for rapid nuclear reactions were existing only for a very short time."

10. *Oh, would you care for a drink?*: By 1956 Gamow had developed a serious drinking problem which was only aggravated as a result of his divorce and his move from George Washington University to the University of Colorado. Vera Rubin, who was his PhD student and later became a distinguished astronomer, recalled conversations with Gamow during which he "finished ½ a bottle of something." See Gingerich (1994, p. 36). Harper (2001, p. 367) describes the years 1956–60 as "the nadir of his life."

11. *your popular book on atomic energy*: In Gamow (1946b, pp. 79–88) he described the new "breath-taking picture of the creation of our Universe" in "an almost instantaneous expansion, of an explosive nature." The starting point of his scenario was a state in which the visible part of the universe "was squeezed into a sphere with a radius . . . ten times the radius of the orbit of the moon!"

12. *my ideas were quite independent of Lemaître's*: There is no reason to disbelieve Gamow, who in his autobiography did not even mention Lemaître and his theory. For his part, Lemaître never referred to Gamow's work on the early universe. The two pioneering cosmologists never exchanged letters or were otherwise in contact.

13. *your famous book*: The book (Gamow 1952a) was a great success. It was reprinted several times and in 1961 it appeared in a new edition that was reprinted in 1970.

14. *I think Einstein liked it*: In a letter to Einstein of 24 September 1946, Gamow summarized his recent idea, namely, "that in order to explain the present relative abundance of chemical elements one must agree that in 'the Days of Creation' the mean density and temp. of the Universe was 10^7 gm/cm^3 and 10^{10} °K." He also suggested that the universe might be in a state of rotation and would therefore have to be described by anisotropic solutions to the cosmological field equations. This, he said, "would have the advantage that, unlike in ordinary simmetrical [*sic*] solutions, the radius of the universe would not go through zero in the beginning of time (even though $\Lambda = 0$)." Einstein dismissed Gamow's idea of a rotating universe and expressed no support for the hypothesis of a hot and dense beginning. However, he did so in a letter of 4 August 1948. Gamow probably mixed up the two letters. See Kragh (1996, pp. 109–110 and 117) and Alpher and Herman (1972, p. 310).

15. *nine out of ten of my ideas are wrong*: The nuclear physicist Edward Teller, who was a collaborator and close friend of Gamow, said: "I'm sorry to say that ninety percent of Gamow's theories were wrong, and it was easy to recognize that they were wrong. But he didn't mind. He was one of those people who had no particular pride in any of his inventions. He would throw out his latest idea and then treat it as a joke!" (quoted in Chernin 1994, p. 794).

16. *His thesis attracted a lot of attention*: On 16 April 1948 *The Washington Post* reported on Alpher's dissertation, comparing his theory of a nuclear origin of the world with the danger of a man-made nuclear destruction of it. Eight days later *Science News Letter* summarized Alpher's (or Gamow's and Alpher's) creation theory as follows: "At the very beginning of everything, the universe had infinite density concentrated in a single point. Then just 300 seconds—five minutes—after the start of everything, there was a rapid expansion and cooling of the primordial matter." At the time Ralph Alpher had started collaborating with Robert Herman, a physicist from the Applied Physics Laboratory, Johns Hopkins University. What is often called Gamow's theory of the early universe was jointly developed from 1948–53 by the three physicists.

17. *This paper of ours is often referred to as the αβγ paper*: The paper is Alpher et al. (1948), which today is considered to be a pioneering paper of early big bang theory.

18. *it was really a joke*: Gamow was the eternal prankster, famous in the physics community for his numerous jokes. Many of them are included in Reines (1972). He revealed the αβγ joke in Gamow (1951, p. 398) (which apparently escaped CCN's attention). The famous physicist Hans Bethe, a friend of Gamow's, happened to review the paper for *Physical Review* and thus knew about the joke, but he just let it pass (Kragh 1996, p. 113). Gamow (1949, p. 369) included a passage on "The neutron-capture theory of the origin of nuclear species recently developed by Alpher, Bethe, Gamow, and Delter," where "Delter" was an author invented for the sake of turning the αβγ theory into an αβγδ theory.

    Mr Tompkins was the hero of Gamow's famous series of popular science books starting in 1939 with *Mr Tompkins in Wonderland*. Tompkins' initials were "C. G. H.," an allusion to the natural constants c, G, and h. In a draft of a paper on proteins and nucleic acids, Gamow included C. G. H. Tompkins as a co-author. Alas, when the paper appeared in the proceedings of the Royal Danish Academy in 1955, it was with Gamow as the sole author. See Gamow (1970, p. 145) and the 1968 AIP interview with George Gamow (see AIP Interviews).

19. *the guys in the Vatican are interested in it*: In 1951 Gamow was informed by an archbishop in Rome that one of his articles on cosmology "was presented to the Holy Father who read it with satisfaction and who looks forward to the publication of your book on 'The Creation of the Universe'" (see Kragh 1996, p. 117).

20. *a lengthy quotation from the pope's address*: Gamow was an atheist and avoided references to religious issues in his scientific as well as popular works. He introduced a paper (Gamow 1952b) on the formation of galaxies by quoting as "an unquestionable truth" several passages from the pope's 1951 address

that fitted perfectly with the Gamow–Alpher–Herman theory. On that address, see Chapter 9. Gamow's views on religion are reviewed in Kragh (1996, pp. 89, 117, 257–258).

21. *the biggest blunder in his life as a physicist*: It is unknown whether Einstein actually said so, but according to Gamow (1970, p. 44), "when I was discussing cosmological problems with Einstein, he remarked that the introduction of the cosmological term was the biggest blunder he ever made in his life." A similar statement appears in Gamow (1956a).

22. *Baade showed that the Hubble constant is much smaller*: Working at the 200-inch Hale telescope, the German-American astronomer Walter Baade had for some time been suspicious of the calibration of the period–luminosity relationship. He argued that the classical population I Cepheids found in globular clusters had a different relationship between period and luminosity than the Cepheids used by Hubble and others to determine the distance to the Andromeda galaxy (Smith 2006). As a result, Hubble had underestimated the size of and distance to Andromeda and overestimated the recession factor. During the 1952 meeting of the International Astronomical Union in Rome, Baade announced that, as a consequence of the required recalibration, the Hubble time had to be increased from 1.8 billion years to 3.6 billion years or more. On Baade's important revision of the cosmic distance scale, see Osterbrock (2001, pp. 162–176) and also Chapter 11.

23. *we live in a hyperbolic universe*: "With the observed value of these two quantities [the mean density of matter and the rate of expansion] one can calculate that the curvature of our Universe is negative, so that space is open and infinite. It bends the way of a western saddle does. The radius of curvature comes out as five billion light-years." (Gamow 1954a, p. 58).

24. *as Hilbert showed a long time ago*: The great German mathematician David Hilbert argued that infinity was a purely mathematical concept without any basis in physical reality. To drive home his point he devised in a lecture of 1924 a highly counterintuitive thought experiment known as "Hilbert's hotel." An account of the weird hotel with its infinity of rooms can be found in the Wikipedia article on "Hilbert's paradox of the Grand Hotel." Gamow was fascinated by the story of Hilbert's hotel, which he first described in a popular book of 1947 and again in his book on the creation of the universe. See Gamow (1947, pp. 16–17) and Gamow (1952a, p. 36). For a history of the origin of Hilbert's hotel and Gamow's use of it, see Kragh (2014b).

25. *Ulam once told me about Gödel's theorem*: See Ulam (1972, pp. 274–275). According to the incompleteness theorem, stated by the logician Kurt Gödel in 1931, it is always possible to construct true mathematical statements that cannot be proved. The Polish-American mathematician Stanislaw Ulam was a close friend of Teller and Gamow. He played an important role in the Manhattan Project and later in the theory of the hydrogen bomb.

26. *what God was doing before he made the world*: See Gamow (1952a, p. 37) and Gamow (1954a, p. 63), where he quoted St Augustine's *Confessions*. However, Augustine thought that before God made the world he made nothing at all.

27. *our expanding universe is the result of an earlier contraction*: In Gamow (1954a, p. 63) he wrote: "We conclude that our Universe has existed for an eternity of time, that until about five billion years ago it was collapsing uniformly from a state of infinite rarefaction; that five billion years ago it arrived at a state of maximum compression in which the density may have been as great as that of the particles packed in the nucleus of an atom . . . , and that the Universe is now on the rebound, dispersing irreversibly toward a state of infinite rarefaction." Gamow (1952a, p. 36) introduced the term "big squeeze" for the state of maximum compression. Today it is mostly referred to as the big crunch.

28. *amounts of dark matter in intergalactic space*: According to Gamow (1956a, p. 144), "If intergalactic space contained matter whose total mass was more than seven times that in the galaxies, we would have to reverse our conclusion and decide that the universe is pulsating."

29. *I have never used the name myself*: In his 1968 AIP interview, Gamow said: "I don't like the word 'big bang'; I don't call it 'big bang', because it is a kind of cliché." (see AIP Interviews). Nonetheless, in a popular article of 1961 he did use the term. See Gamow (1961, p. 104).

30. *a kind of fundamental unit of temperature*: In one of his last papers, received by *Nature* on 22 July 1968 and published three days before his death, Gamow (1968) proposed a series of new units that he suggested would be convenient in cosmology. Among them were "eon" (10^9 years), "hubble" (10^9 light-years), and "inferno" (10^9 K). The length scale currently known as the Hubble length or radius is the distance travelled by light in one Hubble time, $r = cT = c/H$. It is not a fixed unit, but one that increases with the expansion of space.

31. *matter density will exceed the radiation density*: In Gamow's model of 1948 the matter density varied with time as $t^{-3/2}$ and the radiation density as t^{-2}. As he pointed out, it follows that after a certain period of the expansion, what he called the crossover time, the two densities will be equal. He estimated that this would be the case when the universe was about a billion years old and had cooled to about 1000 K, conditions that he thought were necessary for the first formation of galaxies.

32. *it's almost like in the Bible*: In Gamow (1952a, p. 48) he referred in jest to the analogy between Genesis and the early, radiation-dominated universe.

33. *we did most of it collectively*: In fact the prediction of a cosmic microwave background was due to Alpher and Herman alone. In a brief paper of 1948 they calculated the present temperature to about 5 K and the following year

they presented a more detailed calculation. Gamow first mentioned the radiation temperature in 1950, where he suggested 3 K, but without any argument at all. See also Chapter 13. The different approaches followed by Gamow and by Alpher and Herman are analysed in Peebles (2014).

34. *I get a temperature of about 6 K*: Gamow's somewhat obscure derivation suggests that he did not understand the cosmic radiation in the same way as Alpher and Herman. However, he recognized that it was "the residual heat found at present in the universe" and that it was a fossil from the early superdense state. (Gamow 1956b, p. 1731).

35. *possible that Alpher and Herman think differently*: They did. In the years between 1948 and 1958 they made several attempts to promote the big bang theory and to interest astronomers and physicists in looking for the microwave radiation. Nothing came from their efforts. See Alpher and Herman (2001, pp. 118–120) and Alpher (2012).

Their colleague James Follin, with whom they had written an important theoretical paper in 1953, discussed the matter with Allan Sandage in 1954. Sandage (1995, pp. 3–4) recalled: "He said that the George Washington University group was engaged in mounting efforts to observe the birthday of the universe directly. Their plan was to use rockets to observe the highly redshifted Lyman alpha line which they believed was related in some way to their predicted 5 °K relic radiation. This all sounded off the wall to me, not knowing of the 'new cosmology,' having come from a training as a simple classical (observational) cosmologist concerned solely with problems of galaxy distance and redshifts. However, Follin was dead serious, talking Sanskrit to my naivity still dominated by Hubble's ghost."

36. *A Japanese by the name Hayashi*: In 1950 Chushiro Hayashi, at Nanikawa University, Japan, proposed a more sophisticated version of the Gamow–Alpher–Herman theory that led to a present ratio between hydrogen and helium of 6 to 1. Six years later he found a way to produce heavier elements in the early big bang universe, but in amounts that disagreed with observations.

37. *we don't know the amount of helium precisely*: In 1961 the two American astrophysicists Donald Osterbrock and John Rogerson published a study in which they concluded that the composition of the Sun was roughly the same as that of the planetary nebulae, namely, about 32% helium and 64% hydrogen. Referring to Gamow (1949), they suggested that most of the helium "could be at least in part the original abundance of helium from the time the universe formed, for the build-up of elements to helium can be understood without difficulty on the explosive formation picture." See Osterbrock and Rogerson (1961, p. 134). According to Peebles et al. (2009, p. 59), this was "the first well-documented proposal for a relation between the theory and the observational evidence of a fossil from the early universe."

38. *we can account for ninety-nine per cent of all matter*: It was known at the time that only about one per cent of the mass in the universe consists of elements heavier than helium. On the other hand, there was disagreement about the hydrogen–helium composition. For example, Hoyle argued that ninety-nine per cent of all stellar matter was in the form of hydrogen. Gamow and his collaborators started their calculations with either neutrons or a mixture of neutrons and protons. Most of the hydrogen would be formed by the radioactive decay of neutrons, while helium would be formed in collision processes involving protons and neutrons.

39. *the famous mass gap problem*: In order to build up heavier elements, atomic nuclei with mass numbers $A = 5$ and $A = 8$ are needed. The problem was that no such nuclei exist and it proved impossible to bridge the gap by means of other processes that could occur under the conditions of the early hot universe.

40. *a mixed origin of the composition of matter*: Although the theory of Gamow and his collaborators originally aimed at explaining the formation of all elements, he realized that its failure to do so did not amount to a disproof. "I would agree," he wrote, "that the lion's share of the heavy elements may well have been formed later in the hot interior of stars." (Gamow 1954a, p. 62).

41. *cannot explain the formation of galaxies and their different ages*: Referring to this problem, Gamow (1954a, p. 58) concluded: "As far as observations go, the weight of evidence at present is definitely in favor of the idea of an evolving Universe rather than a steady-state one such as envisioned by Bondi, Gold and Hoyle." See also Gamow (1954b) and Kragh (1996, p. 276).

42. *the reddening effect found by Stebbins and Whitford*: On this effect and Gamow's use of it, see Gamow (1954a, pp. 58–59). See also Chapter 9 and Kragh (1996, pp. 276–279). Gamow (1952a, p. 40) characterized the steady-state theory as "artificial and unreal."

43. *Bohr's suggestion of energy non-conservation*: In his 1968 AIP interview, Gamow said about matter creation in the steady-state theory: "It's perfectly logical. Matter is not conserved—okay. Bohr wanted energy not to be conserved. Turned out to be wrong, but entropy is not conserved; entropy increases— so what?" (see AIP Interviews). For Bohr, Landau, and Gamow on energy non-conservation in the early 1930s, see Kragh (1996, pp. 86–87). In his *Constitution of Atomic Nuclei and Radioactivity*, a textbook of 1931, Gamow dealt with the then mysterious spectrum of beta radioactivity. He approved Bohr's idea that "the idea of energy and its conservation fails in dealing with processes involving the emission or capture of nuclear electrons" (p. 56).

   Although Bohr's idea of energy non-conservation turned out to be wrong, it was revived in 1936 by Paul Dirac, who thought that the problems of quantum electrodynamics necessitated abandoning detailed energy conservation. Dirac's proposal was short-lived, but for a while it was supported by the nuclear physicist Rudolf Peierls. See Kragh (1990, pp. 170–173).

44. *a BBC radio debate with Hoyle*: Alpher and Herman (1990, p. 135) mistakenly believed that Gamow participated in the BBC programme with Hoyle. Perhaps they misunderstood something Gamow had told them, or perhaps Gamow just told the story as one of his perpetual jokes.

45. *discussions with Hoyle this summer in California*: According to Hoyle (1981), Gamow had just purchased "an enormous white Cadillac convertible." He recalled "George driving me around in his white Cadillac, explaining his conviction that the Universe must have a microwave background, and . . . my telling George that it was impossible for the Universe to have a microwave background with a temperature as high as he was claiming [higher than 3 K]."

46. *theories that are not founded on Einstein's equations*: The aim of cosmology, Gamow (1949, p. 367) emphasized, was "to see whether or not the problems of cosmology and cosmogony can be understood entirely on the basis of the 'old fashioned' general theory of relativity in its original form proposed by Einstein." His approach to cosmology was similar to that of Tolman, whose textbook (Tolman 1934) was his standard reference book. For Gamow's engineering style of cosmological research, see Kragh (2005).

47. *I consider myself a factual cosmologist*: In an address to a conference in Denver, Gamow distinguished between what he called factual and postulatory cosmology. His own approach, which belonged to the first class, was "to accept the physically established laws governing matter and radiation and look for cosmological models which are derived on the basis of these laws and are consistent with astronomical observations." Moreover, it was similar to "when an engineer wants to design an automobile, a jet plane, or a spaceship, [and] starts with the well-known physical and chemical properties of the materials he uses and looks for the arrangement of these materials which would satisfy his purposes." See Kragh (1996, p. 136).

48. *the postulatory cosmologists in England*: A reference principally to the steady-state theory, and possibly also to Milne's theory of kinematic cosmology. Gamow may also have thought of Dirac's idea of a varying gravitational constant.

49. *Already Kant*: In his classic *Critique of Pure Reason* from 1781, Immanuel Kant analysed the concept of the universe from a philosophical point of view, concluding that cosmology was inherently contradictory and therefore could not possibly attain the status of a science. See also the interview with Lemaître in Chapter 7.

50. *no absolute creation*: Although the Gamow–Alpher–Herman big bang theory was often described as a creation theory, none of the three physicists considered it as such. Gamow, however, frequently used the term, as in his book *The Creation of the Universe*. In the second printing he explained that the term should be understood "not 'making something shapely out of

shapelessness,' as, for example, in the phrase 'the latest creation of Parisian fashion' " Gamow (1952a, preface).

51. *a disappointment, I would even say a failure*: On the reception of the theory, see Kragh (1996, pp. 135–141). The status of the exploding universe theory changed dramatically in 1965 with the discovery of the cosmic microwave background and the emergence of modern big bang cosmology. However, the early contributions of Gamow and his collaborators were not generally appreciated and are still not widely known in the community of astrophysicists and cosmologists. For a detailed technical analysis, written by one of the pioneers of modern big bang cosmology, see Peebles (2014).

52. *people like Fermi and Wigner*: In the years around 1950 several leading American nuclear physicists, including Nobel laureates Enrico Fermi and Eugene Wigner, studied nuclear reactions in the early big bang universe as suggested by Gamow and his collaborators.

53. *Chandrasekhar considers it too speculative*: According to Alpher and Herman (1988, p. 26): "Gamow was concerned that the model being developed would be considered too speculative for the *Astrophysical Journal*, and in particular would not be acceptable to Subrahmanyan Chandrasekhar, one of its editors." Chandrasekhar was an outstanding theoretical astrophysicist who in 1983 received the Nobel Prize in Physics. He served as managing editor of the prestigious *Astrophysical Journal* from 1952 to 1971.

54. *a conference in Belgium*: The conference on nuclear astrophysics took place in Liège from 10–12 September 1953. Apart from Alpher and Herman, their collaborator James Follin also participated.

55. *a summer school in Ann Arbor*: On the University of Michigan's 1953 summer school, see Gingerich (1994). See also Harper et al. (1997, p. 98). The young astrophysicist Donald Osterbrock participated in the summer school and listened to Gamow's lectures on cosmology. "He was always humorous, but with plenty of good ideas," he recalled. "By that time in his life he was a fairly heavy drinker, but it never seemed to mar his thoughts nor his lectures." Peebles et al. (2009, p. 88).

56. *Alpher and Herman have gone into industry*: In the mid-1950s Alpher and Herman took up research positions at General Electric and General Motors, respectively. Although they continued for a period working with cosmological questions, they were effectively cut off from the academic community of cosmologists and astrophysicists.

57. *I'm a bit on the side-lines*: Between 1952 and 1964 Gamow only published a few articles on cosmology, most of them in popular journals and magazines. However, he did keep abreast with the development, both in theory and in observation.

58. *something called the Kalinga Prize*: The Kalinga Prize, consisting of a medal and 20,000 US dollars, was created in 1952 as a UNESCO prize for exceptional skill in presenting science to lay people. Gamow was the fifth recipient of the prize, which in 1967 was awarded to Hoyle. Other early Kalinga Prize

recipients include Louis de Broglie (1952), Bertrand Russell (1957), and Paul Couderc (1966).

59. *I often use popular writings*: Gamow (1970, p. 155) started writing popular works "probably because I love to see things in a clear and simple way, trying to simplify them for myself."

60. *a talk that I didn't understand a word of*: For the meeting in Warsaw and Cracow (which took place in 1938 and not in 1939), see the interview with Eddington in Chapter 9. Gamow did not give a talk, but commented critically on Eddington's presentation.

61. *may have harmed my scientific reputation*: "Gamow committed an unforgivable sin," wrote the physicist Wolfgang Yourgrau, who was his colleague in Denver. "He wrote popular books on physics, biology and cosmology. . . . Most scientists do not fancy the oversimplifying, popularizing of our science . . . [which] is tantamount to a cheapening of the sacred rituals of our profession . . . many of us considered him washed up, a has-been, an intemperate member of our holy order." (Yourgrau 1970, p. 39).

62. *working on a science fiction novel*: Fred Hoyle published *The Black Cloud* in 1957. The book was well received and translated into many languages. Today it is considered a classic in the science fiction genre.

63. *deemed unacceptable for political reasons*: During the first decade after World War II cosmological models with a finite past or of a finite size were *theoria non grata* in the Soviet Union, where they were held to violate the principles of dialectical materialism. See Kragh (2013e) for a survey of the development. The eminent Russian physicist and cosmologist Igor Novikov recalled that, "at the beginning of the 1950s the theory of an expanding, indeed an evolving, universe with the beginning of time at some finite period ago was practically forbidden in the USSR." (Peebles et al. 2009, p. 100).

64. *warning of Western scientific decadence*: In a review in *Astrophysical Journal* (**110** (1949): 315–318) Otto Struve quoted a distinguished Russian astronomer: "The Americanized apostate Gamow . . . advances new theories only for the sake of sensation [and] with amazing ease, sometimes even after a few months, discards them in order to propose a new, equally sensational theory." Struve was born in Russia but had worked in the United States since 1921, where he served as director of the Yerkes Observatory from 1939 to 1950.

65. *a firm stand against communism*: As is evident from Gamow's correspondence, he feared and hated communism. In a letter to Bohr of 24 October 1945 he expressed his fear of "the . . . great Deluge comming [*sic*] from the East which is bound to engulf the free man on the Earth." (quoted in Kragh 1996, p. 106). According to his friend Edward Teller, himself a sharp critic of communism, he was violently anti-communist: "He was terribly, terribly unhappy about the Soviets, and Stalin, about the Communist government." (in Harper et al. 1997, p. 125).

11

Fred Hoyle and Hermann Bondi: the Steady-State Theory

Interview conducted at the Hotel Metropole, Brussels, on 14 June 1958. Language: English.

After my interview with George Gamow I felt it necessary to get the opinion of the steady-state protagonists on the other side of the cosmological debate. I spent most of the summer of 1958 in Europe and had arranged to meet with the three founders of the theory, Fred Hoyle, Hermann Bondi, and Thomas Gold, for an after-dinner conversation. They had been invited to participate in the eleventh of the prestigious Solvay conferences, dealing with astrophysics and cosmology, and we consequently met in Brussels (Figure 12). The conference coincided with the opening of the Brussels World Fair, which I had the pleasure of visiting, including the new and spectacular Atomium building. My plan was to have a joint conversation with the three men, but unfortunately Gold, who had come from Boston, was unable to participate because he did not feel well. Hoyle had also come to Brussels from America, in his case from California where he spent most of the year.

In Brussels I had the pleasure of meeting Lemaître, who was kind enough to remember our conversation at Notre Dame twenty years ago. I also exchanged a few words with Dr Sandage, whom I had briefly met when I was in California to interview Hubble. Moreover, I got the opportunity to speak with another of the invited speakers, the Swede Oskar Klein, and to do it in my native language. When I mentioned to him that I had interviewed his compatriot Svante Arrhenius more than forty years ago, he expressed great interest. It turned out that as a young man he had started in physical chemistry with Arrhenius as his supervisor and only subsequently turned to physics, becoming an assistant to Bohr in Copenhagen. What a small world science is! He had recently become interested in cosmology, he told me, but favoured neither the relativistic explosion theory nor the steady-state alternative. To use his own expression, he was an agnostic.

Figure 12 Thomas Gold, Hermann Bondi, and Fred Hoyle at the General Assembly of the International Astronomical Union, Berkeley, 15–24 August 1961.

CCN To lay the ground for this interview, it seems to me that right now we are at a watershed in cosmology, faced with a choice between two completely different views of the universe, one being the relativistic evolution theory and the other the steady-state theory. We live in a fascinating period, and I can't help feeling that it is analogous to the situation in the late sixteenth century when astronomers were involved in an epic struggle over Copernicus' new ideas. We know the outcome of that ancient debate but not of the present one, where the situation is still open and undecided. Now, the evolution theory covers a broad range of theories or models, and I would like to know whether there are also several steady-state theories. What I have in mind is that the Bondi–Gold theory is sometimes presented as an alternative or rival to the Hoyle theory.[1] I wonder if that is correct, or if it's more reasonable to speak of merely two versions of the same theory, or two different approaches?

HB Fred, let me make a comment on this. It is true that we had rather different ideas when we first formulated our theory and that our approaches and preferences for a good cosmological theory continue to differ to some extent. Tommy and I strongly felt that the extrapolatory approach[2] was unsatisfactory, which is why we for-mulated our theory deductively, with the creation of matter being a strict consequence of the perfect cosmological principle.[3] We want a simple, deductive, and sharply testable theory that stands apart from those based on the equations of general relativity. Don't misunder-stand me, Mr Nielsen, Einstein's theory is correct, of course, but we should not assume that it is valid for the universe at large. I take the view that cosmology is more than just applied general relativity,[4] which is not the commonly accepted view.

Fred is more inclined to what we call the extrapolatory approach and he has constructed a theory where the creation of matter follows from a modified form of the cosmological field equations.[5] He accepts the perfect cosmological principle, but does not consider it axiomatic or of quite the same importance as we do. It comes out of his equa-tions. Nor does he place the same emphasis on simplicity and lack of flexibility as we do—do you find this a fair description, Fred? I don't want to misinterpret your ideas.

FH It's fair enough for me. We don't really disagree, it's just that I think there is more empirical power in a field-theoretical formulation, and to me power is not antithetical to flexibility. My formulation—and also the one suggested by McCrea[6]—is better suited to accommo-date new observations and it offers an understanding of the relation-ship between cosmology and subatomic physics that Hermann and Tommy's version does not. We agree that the large-scale structure of the universe is determined by the rate of creation of matter, but surely this must be a parameter that is explicable in terms of quantum mechanics. Well, the important thing is that our different versions lead to the same picture of the universe and to the same predictions. There really aren't two theories, there's only one. In fact, we agreed early on that the three of us should present our view of the universe as a united theory.[7] And this is what it is.

CCN Just one question concerning steady-state and relativity theory. I understand from what you said, Professor Bondi, that although yours and Gold's cosmology does not build on Einstein's theory, you have no doubt about the correctness of general relativity. So, it's only as a cosmologist that you find it insufficient?

HB Pray, don't call me a cosmologist,[8] I'm a physicist! But yes, you're right. I have the greatest admiration for the theory of general relativity. It's a branch of mathematical physics that increasingly attracts me, at times even more than cosmology. I have recently taken an interest in the thorny question of gravitational waves,[9] a problem which is far from clarified and that I want to investigate in more detail. We don't even know whether the radiation exists or not. It is gravitational physics, but with no direct connection to cosmology, as far as I can judge.

CCN I see. Now, this wonderfully ambitious name "perfect cosmological principle," was that your suggestion, Professor Bondi?

HB No, it was Tommy's invention.[10] I found it to be rather presumptuous, but soon accepted it. Tommy can be very persuasive. It's just an extension of the ordinary cosmological principle to four dimensions, but some people have criticized it for being of an a priori nature and used it to label our theory as pseudo-scientific,[11] which is of course ridiculous. The whole point of our theory is that we want to make cosmology more scientific, and I think we have succeeded in raising the scientific standards by offering an alternative to the evolution theory.

FH Oh yes, we definitely have.

CCN With regard to the perfect cosmological principle, I wonder if you were aware, when you discussed these cosmological questions after the war, that somewhat similar views were proposed as early as the 1920s. There were people who discussed not only an eternal and unchanging universe, but also continuous creation of matter.

HB No, we weren't, not that I recall. Does it ring a bell, Fred?

FH Nope.

CCN I was just curious. These ideas were based on a static stellar universe and are today of historical interest only. Yet, there was an American astronomer[12]. . . . No, never mind, forget about it. There's no point in letting historical details divert our attention. As to the present discussion, it seems to be a mixture of strictly scientific issues and issues of what might be called non-scientific or perhaps of a more philosophical nature. There was some years ago an interesting discussion between you and Gerald Whitrow[13] on whether or not cosmology is a proper science, or at least one heading in that direction. Would you care to say something about it, or perhaps you, Professor Hoyle, have an opinion?

FH Of course cosmology is a science. . . . Well, the steady-state theory is, at any rate. If it wasn't it would be a waste of time dealing with it and I would retreat to the safer haven of astrophysics. But I shall leave the comments to Hermann. He was part of the debate and he has a more philosophical head than I do.[14]

HB Well, perhaps I have. It was actually a discussion broadcast on the BBC which was then written up for a philosophical journal, you know, and our discussion was limited to a very general level—it had to be accessible to the philosophers. We both agreed that cosmology is a somewhat immature science, but we had different opinions about whether or not the immaturity would remain. Whitrow seemed to believe that cosmology will always be some kind of hybrid—half science and half philosophy.[15] Look, compare the situation with physics or astronomy. It would be ridiculous to discuss whether physics is a proper science or not—but it would not be ridiculous in psychology or, say, climatology. What psychology and cosmology have in common is that they are both young sciences, they have not *yet* become fully mature.[16] But they are on their way, cosmology certainly is and rapidly so.

CCN Is that the general attitude among physicists and astronomers?

HB I think it is. A couple of years ago when I attended a conference on relativity in Berne,[17] I pointed out that even protagonists with widely different views of the universe agree when it comes to the basic criteria of comparing theories. I'm not saying that philosophical preferences are of no importance at all, but when we have to decide between competing theories scientists of different cosmological schools adopt the same criteria as in any other science. They compare theoretical predictions with reliable observations and rule out those theories that do not pass the tests. Really, that's all there is to it. In a recent lecture in Manchester I emphasized this point in relation to Popper's analysis, just to make sure that the audience realized that modern cosmology is indeed a science and not some kind of airy crackpot speculation.[18]

CCN And you, Professor Hoyle, do you agree?

FH I guess I do, I'm just surprised that one has to philosophize about such matters. We can do our job without the help of philosophers.

HB Which is not to say, Fred, that there are no philosophical problems in cosmology, or perhaps we should call them conceptual problems.

I recently spoke with Whitrow and he pointed out to me some rather disturbing consequences of the steady-state universe[19]—not disturbing in the observational sense, but in a conceptual and philosophical sense.

FH I wonder what those are . . . , hmm.

CCN Wouldn't one of them be the so-called Hilbert's hotel paradox, where one imagines a hotel with an infinite number of rooms? All of the rooms are occupied and still the hotel manager has no problem in accommodating new guests, not even an infinite number of them! One might argue that if the steady-state theory is true, then we live in such a bizarre kind of hotel[20] with the galaxies corresponding to the hotel guests. Can the real world be that bizarre?

FH I don't know what this Hilbert hotel is, Mr Nielsen, but to me it appears as a scholastic thought experiment of no relevance to science. True, there's an infinite number of galaxies in our universe, and so what? But to enter Hermann's discussion with Whitrow, I do agree that astronomers, although they have different views of the universe, share the same basic standards of science and are perfectly able to communicate in a rational and scientific manner. Just as an example, you know that Sandage is no friend of the steady-state universe, and that's to put it mildly, but nonetheless we entered a fruitful collaboration[21] on how to use observable quantities to distinguish between rival cosmological models. Although we have very different preferences, we had no problem in formulating common criteria for testing theories. . . . That was just to back up your comment, Hermann.

CCN Now, I would like to turn to the question of a beginning of the universe, something that the steady-state theory denies or evades but that other theories have no trouble with. Or perhaps they have, but they are forced to assume a beginning. To most people this is the crucial difference between the two rival views of the universe. They are not so much concerned with matter creation, nor with the geometry of space or with the rate of expansion, but they are concerned with whether or not there was a creation. To put it bluntly and naïvely, what's the problem with a universe that came into existence a finite time ago? One that was created?

FH What's the problem!? It's absolutely fundamental to this class of theories that postulate an absolute beginning. It's not creation I'm against, of course, but the singular creation that is conveniently hidden in the remote past and therefore beyond physical explanation—which continual creation is not. That's a crucial difference. Creationists like Lemaître and Gamow postulate a primordial state of the universe some billion years ago and then claim that it evolved into the universe we observe today by some kind of mysterious "big bang." Where's the proof? I'm just asking. Have we found any fossil from the explosion, any evidence at all? I think not . . .

CCN But isn't . . .

FH Let me just continue my tirade, please, there's more to it. Another problem is that the bang itself cannot be explained, not even in principle, for the laws of nature only came into existence with the universe, they cannot have been there before—there was no before, right? We are invited to see the creation event as due to some cause unknown to science—that's not what they say, but it's implicit in the theory. There can be no natural cause for the event, for a cause necessarily precedes the effect. That's why I described the bang as an irrational process[22] outside science. We are simply leaving science behind us and retreating to a view characteristic of mythological thought.[23] It may not be a problem for religious people, but for people with a scientific outlook it should be recognized for what it is—a huge problem. That's my answer to your question, Mr Nielsen. It may be naïve, but it's nonetheless of crucial importance.

Oh, there's more. You know that we have been severely criticized for violating energy conservation, but what about the bang theory? We say that matter is being created continually and gently, in a way that has testable consequences, that doesn't disagree with empirical knowledge,[24] and that can be understood on a physical basis. The creationists postulate a single but *gigantic* violation of the energy law with no possibility of explaining it. Just out of the blue, or rather out of nothing. Why is continual creation a greater sinner than original creation?

CCN Perhaps because original creation doesn't really violate energy conservation? I remember Arthur Haas once told me[25] that one can assume that the total energy of the universe is zero and always has been zero, and that the original creation of matter and energy was compensated by an increase in negative potential energy due to the expansion—or something like it. I think Jordan in Germany suggested more or less the same, although I can't remember when, it may have been before the war.

FH Interesting, I don't know this fellow Haas, but just two days ago I used in my talk a somewhat similar argument[26] to illustrate the plausibility of matter creation in the steady-state theory. It works there, but I don't think it will work in the context of an explosive universe.

CCN Perhaps not, but let me ask, is it really fair to say that Lemaître and Gamow are creationists, to use your label? In America at least "creationists" are associated with some kind of religious fundamentalism, anti-Darwin and all that, and Gamow doesn't seem to . . .

FH I know, George has no ulterior motives and he doesn't really insist on an absolute beginning in a physical sense. But Lemaître, that's a different story; after all he's a Catholic priest. I like him a lot as a person,[27] but his ideas and cosmological theory belong to the past. In his talk here in Brussels he referred explicitly to God, which I find rather inappropriate. Don't you agree, Hermann?

HB Oh yes, I agree, but at least Lemaître has the decency not to use his theory of the primeval atom apologetically. In fact, he claimed that it's neutral with regard to materialism and theism, although to me it doesn't sound very convincing. But others are much worse. You know, Whittaker.

CCN Whittaker? Who's he?[28]

HB Oh, he died a few years ago. He was an eminent mathematician of the old school, but he also did some good work in general relativity. And then, what is more to the point, he was a devout Catholic who shamelessly used cosmology to prove the existence of God! I actually read his book, just to confirm my worst suspicions. And they were confirmed!

CCN Before we leave this topic I would like to know about this term you have already mentioned, Professor Hoyle, the "big bang." It's used from time to time in the United States, but mostly in newspaper and magazine articles, sometimes in radio broadcasts. I understand that it's your invention?

FH It may well be, I used it in a BBC programme[29] of 1949 and then again in the broadcasts that were turned into my book. I needed a word picture that caught the difference between the steady-state theory and the explosive creation-in-the-past theory, and that's what I came up with. It's just a name and not a particularly apt one. I don't use it myself,[30] except sometimes in conversation and informal talks. Raymond has used it,[31] I know, he likes it. That's all there is to it, just a name.

CCN What about you, Professor Bondi, do you like it?

HB No, it's a childish expression—sorry Fred.

FH No problem, I don't care.

CCN Now, I couldn't help finding it remarkable that here in Brussels, at this prestigious Solvay conference, there are four supporters of the steady-state theory[32] and only a single representative of the explosion theory. I wonder if this may be taken as an indication of an increased recognition of the strength of steady-state cosmology—that it's on its way to become the accepted view of the universe?

HB No, unfortunately it's not. You shouldn't put too much into the composition of this meeting. I don't know who invited the participants, or for what reason, but it is a little surprising that Gamow was not invited,[33] for example, or Jordan,[34] although that may be less surprising—you know, given his past. We may be dominating this conference, or so it may seem, but in reality we are still the underdogs. We are a minority constantly fighting a majority of opponents or sceptics. Take Sandage and Ryle, for example, they may not be advocates of Lemaître's or Gamow's explosion universe but they definitely are against the steady-state theory. The same is the case with Baade, I think, so at best this meeting reflects the lack of consensus in cosmology.

CCN Yes, and Sandage and his group at Mount Palomar have measured a large number of galaxies and arrived at the conclusion, if I've got it right, that the data are incompatible with the steady-state theory,[35] So is your cosmological model now shot down by observations?

FH By no means, although that's how some people would like to have it. But the uncertainty in the method is great, and probably greater than Sandage imagines, which is sort of demonstrated by the more reliable data that were published last year. I admit that measurements of the deceleration don't agree with the steady-state prediction, but they don't clearly disagree with it either. It's an important test, but so far it is inconclusive,[36] at least to my mind. We need sharper methods to settle the question. In fact, I have come up with one possibility,[37] a rather clever one which is based on measurements of the redshifts and angular diameters of clusters of galaxies. I will discuss it at the meeting in Paris, but so far it is uncertain if it can produce more conclusive results than the other method. We'll see. What do you think, Hermann?

HB I have not followed this matter in any detail but would like to add a remark of a methodological nature. You see, Mr Nielsen, the steady-state theory yields precise predictions such as a definite value for the deceleration of the expansion of space, whereas the situation in the class of relativistic evolution theories is completely different. This kind of cosmology does not lead to sharp predictions but to a whole spectrum of predictions, and for this reason it is not as easily testable, and far from as falsifiable, as our theory is. It's a kind of unfair competition,[38] really, at least from the point of view of methodology. As Popper has so cogently argued, falsifiability is the hallmark of scientific theory. The steady-state theory lives a far more dangerous life than the evolution theories, don't you agree, Fred?

FH Absolutely, you know I do. This is something we have often pointed out. Fortunately, in the end the decision is left to nature, and she may not care for either methodology or sociology.

CCN No, a methodologically appealing theory may not be nature's choice. In addition to optical astronomy, more recently radioastronomy has entered significantly into the context of cosmology,[39] which is an issue that I'm eager to know your opinion about. I have the impression that with regard to the testing of cosmological models there are some parallels between the two kinds of astronomy. Is my impression correct?

FH Oh yes, definitely. The short version is that Martin—that's Martin Ryle, my astronomy colleague in Cambridge—first announced that his measurements of discrete radio sources contradicted the steady-state theory, but then we successfully questioned the quality and interpretation of the data. Tommy did that,[40] and Hermann too, I was a little slow in following it up. And then we unexpectedly received help from the Sydney astronomers,[41] whose data further undermined Martin's conclusion, although they didn't precisely correspond to what we expected. That's the short version.

The present situation is muddled, it definitely is, but at least we know that Martin's announcement was premature—and more than that, it was simply unfounded. For a few days ago we listened to Lovell's balanced review,[42] the essence of which was that present radioastronomical surveys are unable to distinguish between rival cosmological models. There is not as yet any verdict, or perhaps one should say that the case is not proven.

HB You're right, Fred, but perhaps Mr Nielsen was referring to the methodological parallels between the two methods. In any case, they are quite striking. The method based on number counts of radio sources[43] is able to conclusively test the steady-state theory, or rather to prove that it's wrong. But it is not able to do the same for the evolution theories. Even should the data happen to agree with the steady-state prediction, it wouldn't mean a refutation of the other class of cosmological models. So this is one more example of the methodological asymmetry, that our theory is living a dangerous life.

CCN So, is it possible to conclude anything at all from the surveys based on radioastronomy?

FH Martin says yes, we say no—and for the moment it seems that we are right. Frankly, Martin is not as neutral and objective in his selection and interpretation of data as one might wish him to be. It's all too clear that he *wants* to disprove our theory[44] and that he has his conclusion ready in advance of the data that are supposed to justify it. We will soon know more, perhaps already by next month when I'm going to Paris to a symposium[45] on radioastronomy. Martin will be there and Mills will speak for the Sydney group. It will be interesting to see if they can agree, or perhaps they will only agree to disagree.

CCN Given the uncertainty of both optical and radioastronomical surveys, can the chemical composition of the universe be of any help? What I mean is, Gamow considers himself a kind of nuclear archaeologist who infers what the cosmological past was like from the abundance of elements that we observe at present, and he says that the large amount of helium is evidence for the hypothesis that you have labelled the big bang. I assume you disagree?

FH Of course we disagree. I'm myself a nuclear archaeologist, to use your apt phrase, but of a different school, and I've recently done a lot of work together with my esteemed colleagues on the synthesis of elements in various types of stars.[46] I actually gave a report on our work here in Brussels just two days ago. Suffice to say that we are able to account for the distribution of elements without assuming a hot beginning of the universe—the stars are hot enough and they

are sufficient to do the job. I'm not saying that this provides solid evidence for the steady-state theory, but at least it's consistent with it
and it weakens the creation-in-the-past alternative, although it does-
n't disprove it. That's the situation in a nutshell. Besides, we don't
know the amount of helium very well, so George is basing his optimism on a quantity that hasn't been measured.

HB Quite so, and it may be worth pointing out that here we have yet
another instance of what we called unfair competition. We have no
hot primordial universe at our disposal, so we are forced to account
for the synthesis of *all* the elements by means of processes going on
now, including helium. Our theory is a kind of cosmological version of Lyell's uniformitarian geology.[47] You know, in the nineteenth
century Lyell dismissed the biblical flood as a valid explanation in
geology, and we dismiss the primordial bang as a valid explanation in
cosmology. That's a nice analogy, I like it.

The point is that Gamow and his associates, or Lemaître for that
matter, have two kinds of nuclear ovens at their disposal, not only
the hot early universe but also the stars. So they have a much greater
freedom to account for the elements, and in this respect their theory is close to being unfalsifiable while ours is not. Fred is right, the
recent work on the formation of elements has not disproved the explosion theory, but it is nonetheless a triumph of the steady-state[48]
point of view.

FH Yes, and then our theory of element formation is of course valid in
its own right, which is even more important to me. Should it turn
out that our world is not in a steady state—God forbid!—then it's
nonetheless a breakthrough in the understanding of stellar nucleosynthesis. Oppenheimer admitted as much,[49] and he's no friend of
our view of the cosmos. He was quite clear about that.

CCN All this is a bit depressing. I now realize that I had, in my naïvety,
thought that this big question of the correct picture of the universe
was about to be solved, but instead I discover that the picture has
only become murkier, apparently without any hope of solution in
the near future. I'm confused. But there may be another possibility,
for Gamow told me that he places some hope in the detection of an
electromagnetic radiation that emerged in the hot era of the early
universe and then cooled as the universe expands. I may have got

it wrong, but it was something like that. I also remember, or I think I remember, that this was something he had discussed with you, Professor Hoyle?

FH Oh yes, it was one among many topics we discussed when I met him two years ago. Even earlier, shortly after we introduced the steady-state theory, I considered a radiation background in connection with Lemaître's theory. I knew that a Canadian astrophysicist had identified a spectral line as due to an excitation temperature of 1 K and realized that it corresponded to a very weak background radiation.[50] But I didn't pay much attention to it and it cannot be the same radiation that Gamow and Alpher have suggested. Their background is much warmer than the one I have in mind, which can be 1 K at most. So long as radiation of the kind suggested by the explosion model has not been detected, we can safely ignore it. Sorry, but that won't help either.

CCN Too bad. What then about the change in the cosmological time-scale? I understand that Baade's work of 1952 was the one that paved the way for a solution of the paradox of the young universe. But, I wonder, did he publish his result?

FH No, that was only a couple of years later. He announced it orally at the Rome meeting of the International Union and it was mentioned briefly in the transactions of the meeting—just a single line about the revision of the time-scale. I know all about it, for I happened to act as secretary for the Commission of Extragalactic Nebulae and thus was responsible for the minute. In fact, other astronomers quickly plagiarized his discovery,[51] pretending that it was theirs.

CCN Really? Who would that be?

FH Well, you know it, Hermann. I don't want to gossip, but it's hardly a secret any longer that Shapley was the culprit, or one of them. Not only did he arrange an article in the *New York Times* without mentioning Baade, he also wrote a paper in which he presented the discovery as were it his own. Baade was not pleased,[52] I can tell you that. Not pleased at all.

CCN Interesting! But apparently Baade didn't relate the new galactic distances to problems of cosmology. What I mean is, he didn't look into the consequences for the various cosmological models. Or am I wrong?

FH You are not wrong. I know Walter very well, he's a great astronomer but he has no respect at all for theoretical cosmology.[53] He always says that what matters is to get better data on stars and galaxies, whereas he considers discussions of the universe as a whole to be premature or something that should be given low priority. In fact, he shares some of the sentiments of earlier critics of cosmology, and he simply dislikes the cosmological principle[54] which he tends to see as more ideology than science.

HB And he considers the perfect cosmological principle to be even worse, of course.

FH Yes, of course. Nonetheless, he's a wonderful astronomer.

CCN This is very interesting. Now, let me return to the new insight that the age of the universe—well, I mean the Hubble time—may be as high as seven billion or even ten billion years. Then the time-scale difficulty has disappeared and with it also the main argument of the steady-state theory against the evolution models. Not that the new time-scale proves this class of models, but it does weaken the steady-state alternative, doesn't it?

FH That's what people keep telling me, but they're wrong. First, the short time-scale was never our main argument, for it was just one argument among many. Second, it's a misunderstanding to believe that the problem has disappeared. The new time-scale has not been narrowed down very precisely, but whatever its value it still conflicts with the ages of the oldest stars,[55] which are more than ten billion years. In my view, a time-scale of the order of seven or eight billion years, and that seems to be what Sandage favours, doesn't change much in the balance between the two theories of the universe. It neither proves nor disproves any of them, and it doesn't really make the picture less murky.

CCN Well, then I have run out of ideas, I think. Perhaps we should . . . , no, I forgot one thing. Professor Bondi, in your remarks concerning the methodological advantages of the steady-state theory you referred briefly to Popper, who also turned up in your discussion with Whitrow.[56] I assume you referred to the philosopher by that name, Karl Popper, isn't it?

HB Yes, he was born in Vienna, just like I was—and Tommy too. He's now at the London School of Economics. It was only after I moved to King's College in London that I came to know him personally, but I knew of his philosophical ideas earlier; they are very attractive both when for science and society. Are you aware of his great work on the logic of science? It's in German, but next year it will appear in an English edition.

CCN I'm aware of it, yes, but I must admit that I haven't read it. Philosophical treatises are not my favourite literature. But perhaps I should read Popper's book, it may be more to my taste.

HB You should, I studied it in about 1950, and I instantly liked it. What strikes me is that Popper, contrary to most philosophers, thinks like a physicist and speaks the language of physicists.[57] It takes some translation from his more logical formulations, of course, but it's the same language. He proudly showed me a letter that Einstein wrote him shortly after the book came out in the thirties. . . . Einstein was quite enthusiastic[58] and so am I. Well, I informed him about the situation in cosmology, pointing out how nicely his point about falsifiability fits with the steady-state theory and how badly it compares with the relativistic theories. I'm not sure I have convinced him, though, but he told me in no uncertain words that he considers the bang-in-the-past hypothesis to be plainly unscientific.

On the other hand, while he likes the idea of continual creation of matter, he doesn't accept its foundation in the perfect cosmological principle[59]. . . . Anyway, he really is a great philosopher[60] and about the only one who working scientists need to pay attention to. In my Joule lecture I did some advertising of his ideas[61] and their relevance for cosmological research. They deserve to be advertised.

CCN This is very interesting, for it's not often that philosophers make an impact on science, as far as I know. Now I know the opinion of Popper and it just crossed my mind, what about Einstein? Do you know how he felt about the theory of the steady-state universe?

FH No, I don't, do you Hermann?

HB No, he probably disliked it, if he knew about it at all.[62] It's of no importance.

CCN Yes, I just wondered. I once had a conversation with him, but that was a long time ago, in Berlin. But it's time to end our conversation, I believe. You are very busy and I understand that Brussels is only one of your destinations this year. Moscow will be next, right?

FH I'll go to Moscow, yes, but only after I have been to Paris. Cosmology is practically non-existent in the Soviet Union,[63] which in part is for political and ideological reasons, so the topic may not be high on the agenda. I've been told that finite-age models are out of the question, but also that continuous creation of matter is considered illegitimate. That's funny, for what's left, then?

CCN Funny it is, or perhaps one could find another and more appropriate word. . . . Well, goodnight to you, gentlemen!

Notes

Fred Hoyle was born on 24 June 1915 in Bingley, Yorkshire, England and died on 20 August 2001 in Bournemouth, Dorset, England. Young Hoyle did not like school, but he did like science. Having left Bingley Grammar, he was admitted to Cambridge University in 1933 on a scholarship to study theoretical physics. As a graduate student he specialized in nuclear physics and quantum electrodynamics, but in 1939 he decided to switch to astronomy and astrophysics. He did that successfully and vigorously. Between 1939 and 1947—a period during which he was also much occupied with war-related research—he wrote 25 astronomical research papers. In a pioneering work on nuclear astrophysics from 1946 he showed how the abundance of middle-mass elements in the stars could be understood in terms of equilibrium nuclear processes. For the next two decades he continued to investigate stellar nucleosynthesis, a field in which he was a recognized authority. For example, in 1953 he predicted that the synthesis of carbon from three alpha particles necessitated a resonance level in the carbon nucleus at 7.7 MeV. Experiments made shortly later proved him right. In 1957, together with William Fowler and Margaret and Geoffrey Burbidge, he produced a comprehensive theory of element synthesis in stars, a classic known as the B^2HF theory.

In 1947, as a result of discussions with Thomas Gold and Hermann Bondi, Hoyle became seriously interested in cosmology (see Figure 12). The result was the steady-state theory of the universe, which he defended against the theory of the "big bang," a name he coined in 1949. Rather than abandoning the steady-state model after the discovery of the cosmic background radiation in 1965, he suggested that the radiation was not of cosmological origin and that it could be accommodated within a revised version of the steady-state model. He resisted the big bang theory until the end of his life. In the 1990s, together with Jayant Narlikar and G. Burbidge, he proposed as an alternative a "quasi-steady-state model" that avoided the unacceptable feature of a universe with a beginning in time. Although not bowing to what he saw as the big bang dogma, in the

1960s he made calculations of the formation of light elements under the physical conditions of a big bang. Ever the controversialist, in the 1970s and 1980s he collaborated with Chandra Wickramasinghe on a series of books and articles in which they argued that the chemistry of life was to be found in outer space. In addition to his numerous scientific publications, Hoyle also wrote books for the general public, some of them science fiction novels and others popular books on astronomy.

Biographical sources: Mitton (2005); Gregory (2005); Hoyle (1994). Interview of 15 August 1989 in Lightman and Brawer (1990, pp. 51–66).

Hermann Bondi was born in Vienna, Austria on 1 November 1919 and died in Cambridge, England, on 10 September 2005. He belonged to a Jewish family which had moved from Germany to Vienna in 1884. As a teenager he demonstrated a remarkable talent for mathematics, and in 1937 he was admitted as a foreign student at Trinity College, Cambridge. His studies were interrupted by the war, which caused him to be interned in a camp in Canada for more than a year. Upon his return to England, he was drawn into military research. In 1946 he became a British subject. Together with his friends Fred Hoyle and Thomas Gold—another Jewish-Austrian internee—he developed the ideas that in 1948 resulted in two versions of the steady-state theory of the universe, one due to Hoyle and the other to Bondi and Gold (see Figure 12). Four years later he published *Cosmology*, a book that for a decade or more was the standard textbook on the subject. While defending the steady-state theory methodologically, he contributed relatively little to it, and when the theory ran into serious troubles in the mid-1960s he quietly left cosmology as a research field. Instead he concentrated on Einstein's general theory of relativity, a branch of physics he had first examined in 1947 and which became his primary area of research in his later career. In a series of papers from 1959 to 1988 he conclusively showed that general relativity predicts gravitational waves in a physical (and not merely mathematical) sense.

Bondi had an interest in the philosophy of science and was an early follower of Karl Popper, whose philosophy he greatly admired. In the 1950s he used Popper's criteria of science to argue that the steady-state theory was methodologically superior to the relativistic evolution theories. As an able organizer and administrator, and a believer in the social value of science, from the late 1960s he became increasingly involved in science policy and public service generally, both on a national and an international level. He was instrumental in the establishment of the European Space Research Organization (ESRO, the predecessor of ESA, the European Space Agency), served as chief scientific advisor to the British government, and was chairman of the Natural Environment Research Council. Bondi was an atheist and for a period president of the British Humanist Association, an atheist organization.

Biographical sources: Bondi (1990); Roxburgh (2007); Kragh (2008c). AIP interview of 1978 (see AIP Interviews).

1. *an alternative or rival to the Hoyle theory*: The two founding papers on the steady-state theory were received by the *Monthly Notices of the Royal Astronomical Society* on 14 July 1948 (Bondi and Gold) and 5 August 1948 (Hoyle). In spite of leading to the same results, they differed significantly. Bondi and Gold (1948, p. 269) concluded that Hoyle's version was "unsatisfactory and unacceptable." They had "no hesitation in rejecting Hoyle's theory, although it is the first and at the moment only field theory formulation of the hypothesis of continuous creation of matter." A few years later Bondi (1952, p. 155) adopted a more conciliatory attitude, calling the difference between the two versions "largely a matter of taste." And in his 1978 AIP interview he recalled: "I don't think we ever thought the world models were different. The depth of the split was never very great." (see AIP Interviews).

2. *the extrapolatory approach*: What Bondi (1948, p. 107) called the extrapolatory approach was the extrapolation from laboratory physics "to form a comprehensive mathematical theory of relativity first and consider its cosmological implications only afterwards." He contrasted it with the deductive approach, according to which "one may try to postulate a rigorous cosmological principle and attempt to derive the corresponding theory of relativity and gravitation afterwards." The Bondi–Gold theory was methodologically indebted to Milne's cosmology, while Hoyle's was not. In his widely read textbook *Cosmology*, Bondi (1952, pp. 1–8) elaborated on the relative merits of the two attitudes to cosmology. Rather than considering cosmology "a minor branch of general relativity," to him it was "the most fundamental of the physical sciences, the proper starting point of all scientific considerations."

3. *a strict consequence of the perfect cosmological principle*: Bondi and Gold introduced the perfect cosmological principle in their 1948 paper as the requirement that the universe must be homogeneous and unchanging at a the large scale. They expressed great confidence in the principle, stating that without it cosmology would not be truly scientific. If results from general relativity conflicted with the principle, they were willing to reject the results. Yet, as they cautiously pointed out, "we shall never disregard any direct observational or experimental evidence."

    It follows from the perfect cosmological principle that the mean density of matter ρ must be independent of time and, as a consequence of the expansion of the universe, matter must be continually created. Bondi and Gold showed that the rate of creation was given by $3\rho H \cong 10^{-43}$ g/cm^3/s, where H is the Hubble constant. See also Bondi (1952, pp. 141–144). Whereas H is decreasing with cosmic time in relativistic models, where the Hubble time $1/H$ is a measure of the age of the universe, according to the steady-state theory it is a true constant of nature.

4. *cosmology is more than just applied general relativity*: In the preface to *Cosmology*, dated 27 October 1950, Bondi wrote: "I do not regard cosmology as a minor branch of general relativity or as a branch of philosophy or logic."

5. *a modified form of the cosmological field equations*: Hoyle (1948) based his theory on Einstein's equations of 1917, but modified in such a way that the cosmological term was replaced by what he called a creation tensor. As indicated by its name, the effect of the new term was spontaneously to create new matter. Thus, whereas Einstein's equations (with or without the cosmological constant) conserve energy, Hoyle's did not. Hoyle further derived a definite value for the matter density of the universe, namely, $\rho = 3H^2/8\pi G$. The expression happens to be exactly the same as in the Einstein–de Sitter model of 1932, where H and ρ are, however, decreasing in time.

6. *also the one suggested by McCrea*: The Irish-born physicist William Hunter McCrea was favourably disposed to the steady-state theory from its very beginning, and especially to Hoyle's formulation of it. In a paper of 1951 he proposed field equations for the steady-state universe that in a formal sense were identical to Einstein's cosmological equations. Instead of making use of a creation term, as Hoyle had done, he reinterpreted the energy–momentum term as a term corresponding to space being endowed with negative pressure. For details see Kragh (1999).

7. *present our view of the universe as a united theory*: Following a critical discussion of Hoyle's theory at a meeting of the Royal Astronomical Society on 11 March 1949, Bondi intervened: "As the objections which have been brought against Mr Hoyle's theory would appear to apply equally to the work by Mr Gold and myself, I think that the three of us should present a united front." (*Observatory* **69** (1949): 49).

8. *don't call me a cosmologist*: "I always detest being referred to as a cosmologist," Bondi (1990, p. 63) said in his autobiography. "Though I got fame from that subject, it was far from the only field in which I did research." In the period up to about 1970, very few scientists would identify themselves as "cosmologists," a label that typically carried a negative connotation associated with philosophical speculation. They were physicists, astronomers, or mathematicians who *also* did research in cosmology.

9. *thorny question of gravitational waves*: After he moved to King's College, London, in 1954, Bondi built up a strong research school in general relativity. In papers from 1959–62 he and his collaborators gave the first exact proof that general relativity theory predicts gravitational waves.

10. *it was Tommy's invention*: The ideas of the perfect cosmological principle and continual creation of matter were originally suggested by Gold. See Kragh (1996, pp. 173–179). In an interview of 1978 he recalled his discussions with Bondi: "I said, 'Perfect is the only thing we can do because the word perfect, in any case, has a connotation of a temporal kind in our language.' Where did I get that idea from? Parfait? The past tense or something. . . . Somehow

I liked that . . . and I persuaded Bondi to accept it." Gold, 1978 AIP interview (see AIP Interviews).

11. *label our theory as pseudo-scientific*: Herbert Dingle, in particular, accused the perfect cosmological principle of being dogmatic and unscientific. In 1956, alluding to Aristotelian cosmology, he thundered that it had "precisely the same nature as perfectly circular orbits and immutable heavens," that is, it was in principle inviolable. (Dingle 1956, p. 235). His sharp critique was a continuation of the attack he launched in the late 1930s against rationalist tendencies in cosmology and which is mentioned in Chapters 8 and 12. Bondi and Gold wavered somewhat in their attitude to the perfect principle, on occasions presenting it as a priori true. By 1958 at the latest Bondi made it clear that it should be regarded as nothing but a working hypothesis subject to refutation by observation.

12. *there was an American astronomer*: According to William D. MacMillan, a professor of astronomy at the University of Chicago, the universe as a whole had never been essentially different from what it is today. His idea of a classical steady-state universe was shared by several other scientists, including the Nobel laureates Robert Millikan and Walther Nernst. As mentioned in Chapter 2, it had previously been suggested by Arrhenius. For this tradition see Kragh (1996, pp. 143–160).

13. *an interesting discussion between you and Gerald Whitrow*: On this discussion, highly instructive from both a scientific and a philosophical point of view, see Whitrow and Bondi (1954) and Kragh (1996, pp. 233–236).

14. *he has a more philosophical head than I do*: "Tommy and I were probably a good deal more philosophically minded than Fred." Bondi, 1978 AIP interview (see AIP Interviews).

15. *half science and half philosophy*: Whitrow concluded that physical cosmology was "rooted equally in philosophy and in empirical knowledge, as were the most ancient cosmologies . . . [It] has been, still is and, from its very nature, will, I believe, remain a border-line subject between the special sciences and philosophy." (Whitrow and Bondi 1954, p. 277).

16. *they have not yet become fully mature*: According to Bondi in his discussion with Whitrow: "In recent sciences, such as virus research or psychology, unanimity is absent. What is essential is only that there should be unanimity about the means of deciding what is correct and what is incorrect, that the yardstick of experiment should be universally accepted." (Whitrow and Bondi 1954, p. 278).

17. *I attended a conference on relativity in Berne*: This was a jubilee conference, celebrating the fiftieth anniversary of Einstein's introduction of the special theory of relativity. "I should like to draw attention to how scientific cosmology has become," Bondi (1956, p. 152) said. "The cosmological papers today have all dealt with empirical tests of cosmological theories and nobody has referred to how satisfying or beautiful or how logical this or that theory is."

18. *not some kind of airy crackpot speculation*: In his Joule Memorial Lecture of 1958, Bondi dealt with the nature and status of cosmology. He concluded that, "our theories are not cranky invulnerable speculations but they are proper scientific theories which suggest experiments by which they may be shot down." (Bondi 1958–1959, p. 71).

19. *some rather disturbing consequences of the steady-state universe*: Whitrow (1959, pp. 138–141) showed that it followed from the steady-state theory that the universe contains infinitely many sub-universes which are causally disconnected and therefore beyond empirical recognition even in principle. In other words, the theory leads to what in later cosmology would be called a multiverse: "There will be galaxies which are not accessible to observation at any stage in their history." Whitrow felt the consequence to be so bizarre "from the point of view of the theory of knowledge" that he doubted that the steady-state theory could be a candidate for the real universe.

20. *we live in such a bizarre kind of hotel*: Several years after the cosmological controversy had ended the philosopher N. W. Boyce (1972) argued that there is a perfect analogy between the steady-state universe and the paradoxical Hilbert's hotel. For this hotel, see Kragh (2014b) and also Chapter 10.

21. *we entered a fruitful collaboration*: A reference to Hoyle and Sandage (1956), in which the two astronomers formulated the basic rules for distinguishing between cosmological models by means of the values of the Hubble constant and the so-called deceleration parameter, an observable quantity related to the geometry of space. Sandage later acknowledged Hoyle's role in the project: "I didn't know the equation for the space curvature . . . until Fred Hoyle came over from England and gave a course to the students at Caltech . . . Hoyle's understanding of everything and teaching it to the students was the cornerstone from which I then began to work." Interview in Lightman and Brawer (1990, p. 76).

22. *the bang as an irrational process*: In his controversial and widely read *The Nature of the Universe*, Hoyle (1950, p. 124) wrote about "the big bang assumption . . . [which] is an irrational process that cannot be described in scientific terms." The charge of irrationality appeared in the American edition, but not in the simultaneously published English edition.

23. *a view characteristic of mythological thought*: According to Hoyle (1955, p. 351), the notion of a temporal beginning of the universe presupposed arbitrarily chosen initial conditions. "This procedure," he said, "is quite characteristic of the outlook of primitive peoples, who in attempting to explain the local behaviour of the physical world are obliged in their ignorance of the laws of physics to have recourse to arbitrary starting conditions."

24. *doesn't disagree with empirical knowledge*: Hoyle, Bondi, and Gold emphasized that there was no experimental reason to dismiss a creation of matter as tiny as given by the figure of 10^{-43} g/cm^3/s, or the formation of three new hydrogen atoms per cubic metre per million years. Hoyle (1950, p. 125) expressed the

creation rate as "no more than the creation of one atom in the course of about a year in a volume equal to that of a moderate-sized skyscraper."

25. *Arthur Haas once told me*: See Haas (1936) and, for Pascual Jordan's similar argument of 1938, Kragh (2004, pp. 181–182). See also Chapter 7. After the standard big bang model had been firmly established in the 1970s, the Haas–Jordan idea of a universe with zero net energy played a role in attempts to explain the birth of the universe as a result of vacuum quantum fluctuations.

26. *a somewhat similar argument*: "The process of creation can accordingly be thought of as involving no energy expenditure—a particle is created at a negative potential that compensates for its rest mass." (Hoyle 1958, p. 57). His argument at the Solvay conference was limited to the creation of particles characteristic of the steady-state theory.

27. *I like him a lot as a person*: Hoyle got to know Lemaître well from a conference on stellar populations that took place in May 1957 in a renaissance villa in the Vatican garden. Apart from Hoyle and Lemaître, other participants included Walter Baade, Otto Heckmann, Allan Sandage, Jan Oort, and William Fowler. See photographs in Osterbrock (2001, p. 194) and Struve and Zebergs (1962, p. 281). After the end of the conference Hoyle, his wife Barbara, and Lemaître went on a two-week drive from Rome to Naples and then northwards to Switzerland. In spite of their different views on cosmology and religion, they got along very well. Hoyle (1994, p. 301) later described Lemaître as "a round, solid man, full of jokes and laughter."

28. *Whittaker? Who's he?*: The English mathematician Edmund Taylor Whittaker was influential as a writer of textbooks and also had a deep interest in the history of physics. He converted to Catholicism in 1930. The book Bondi referred to was probably *Space and Spirit* (Whittaker 1946), a sustained attempt to integrate theological and cosmological thought. See Kragh (2004, pp. 193–197).

29. *I used it in a BBC programme*: On 28 March 1949 Hoyle gave a BBC broadcast on the new theory of continual creation in which he coined the name "big bang." Two weeks later the text of the broadcast appeared in *The Listener*, a widely read BBC magazine (Hoyle 1949a). See also Mitton (2005, pp. 125–135). His book *The Nature of the Universe* (Hoyle 1950) was based on a series of five broadcasts he gave in the first months of 1950.

30. *I don't use it myself*: Hoyle ignored the term "big bang" in his publications between 1950 and 1965. As to the origin of the name, he recalled: "I was constantly striving over the radio—where I had no visual aids, nothing except the spoken word—for visual images. And that seemed to be one way of distinguishing between the steady-state and the explosive big bang. And so that was the language I used." Interview in Lightman and Brawer (1990, p. 60). The term only appeared insignificantly in the scientific

literature until the 1970s. For details on its history see Kragh (2013d) and Kragh (2014a).

31. *Raymond has used it*: The British astronomer Raymond Arthur Lyttleton was a friend and collaborator of Hoyle. He much preferred the steady-state theory over relativistic evolution theories. In a popular book of 1956 he compared "the 'big bang' hypothesis" unfavourably with the steady-state theory. See Lyttleton (1956, p. 197).

32. *four supporters of the steady-state theory*: In addition to Hoyle, Bondi, and Gold, the leading steady-state cosmologist William McCrea was also among the invited scientists.

33. *Gamow was not invited*: Gamow very much wanted to join the Solvay conference, but was told that there were no more vacancies. "I was not surprised (though somewhat disappointed) about the outcome, since I was an opponent of the steady-state theory," he recalled in his autobiography (Gamow 1970, p. 126). In fact, the Gamow–Alpher–Herman version of the big bang theory was not mentioned at all at the Solvay conference.

34. *or Jordan*: The German physicist Pascual Jordan, one of the founders of quantum mechanics, did important work in cosmology and general relativity in the 1950s. However, during the Third Reich he was a member of the Nazi party (Schucking 1999). Having passed a denazification process after the war he was fully rehabilitated, and by 1958 he had become a member of the Bundestag, the parliament of West Germany, for the Christian Democratic Party. Nevertheless, his past as a Nazi was not easily forgotten in formerly occupied Belgium.

35. *incompatible with the steady-state theory*: Humason et al. (1956) published redshift and magnitude data for 474 galaxies with redshifts up to $z = 0.2$. For the deceleration parameter they found a best value of $q_0 = 2.5 \pm 1$, whereas the steady-state theory predicted $q_0 = -1$. Sandage was convinced that the results from Mount Palomar effectively ruled out a steady-state universe.

36. *so far it is inconclusive*: Photoelectric measurements from 1957 resulted in a value for q_0 in the interval between 0.5 and 1.5, still incompatible with the steady-state theory but less markedly than previously. Hoyle chose to interpret the discrepancy of the data as a sign of the inconclusive nature of the test. See Kragh (1996, pp. 283–286).

37. *I have come up with one possibility*: Hoyle first discussed the diameter–redshift test in his address to the Paris symposium. See Hoyle (1959) and also Kragh (1996, pp. 286–287). The test only became effective in the 1990s, many years after the steady-state theory had disappeared from the scene.

38. *a kind of unfair competition*: Hoyle (1959, p. 529) expressed the situation as follows: "Relativistic cosmology contains three parameters that must be determined from observation in contrast to the steady-state cosmology, which contains only a single disposable parameter. It is very difficult to

design any stringent test of relativistic cosmology in its most general form, since a limited number of observations can always be explained away by a suitable process of 'parameter fitting.' In the steady-state theory, on the other hand, the single parameter becomes assigned once the value of the Hubble expansion constant is known, and no further degree of freedom them remains in the theory."

39. *radioastronomy has entered significantly into the context of cosmology*: Radio cosmology started around 1954, after it was realized that most radio sources are extragalactic and can therefore be used to gain information of the universe at large. For the early development of radioastronomy and its relevance for cosmology, see Sullivan (1990) and Kragh (1996, pp. 305–317).

40. *Tommy did that*: Gold recalled his early scepticism with regard to Ryle's data, which he suspected were unreliable because of accumulated errors in the intensity measurements: "Hoyle had taken the data very seriously, you see, and I kept saying to Hoyle, 'Don't trust them, there might be lots of errors in this and it can't be taken seriously.' Hoyle said, 'You must take observational data seriously, otherwise you are nowhere.' I said, 'I will take them seriously when I know they're correct and not before.' " (Gold, 1978 AIP interview; see AIP Interviews).

41. *we unexpectedly received help from the Sydney astronomers*: The data collected by Martin Ryle and his collaborators at Cambridge University in the 2C survey of 1955 were contradicted two years later by data from the Southern Hemisphere obtained by Bernard Mills and his group in Sydney. Whereas Ryle suggested that his counts of radio sources were at variance with the steady-state theory, Mills found no clear evidence for or against any particular cosmological model.

42. *Lovell's balanced review*: A. C. Bernard Lovell, a leading British radioastronomer, was the founder and director of the Jodrell Bank Observatory in Cheshire, England. In his Solvay address he argued that, "the only safe conclusion to be drawn from the work [at Cambridge and Sydney] is that, as yet, there are no radio astronomical observations which can influence significantly the existing views on the large scale structure of the universe." (Lovell 1958, p. 201).

43. *The method based on number counts of radio sources*: On the assumption that extragalactic radio sources are distributed uniformly in a static flat space, the number N of sources with an intensity larger than a certain value I will depend on I according to $\log N = -1.5 \log I + $ constant. That is, the two quantities $\log N$ and $\log I$ will exhibit a straight line with slope of -1.5. The steady-state theory predicts that the radio sources will lie beneath the line, whereas the relativistic models do not lead to definite predictions. According to the Sydney group, the major part of the $\log N$–$\log I$ curve had a slope of -1.8, and in 1958 the figure had come down to -1.65.

44. *he wants to disprove our theory*: Hoyle (1990, p. 228) complained: "His programme, which he pursued relentlessly over the years, does not seem to have been directed towards any other end [than disproving the steady-state theory]. There was no question of establishing the correct cosmology, but only of disproving the views of a colleague in the same university." See also Hoyle (1994, pp. 408–410). Partly as a result of their disagreements about the interpretation of the radio data, the relationship between Hoyle and Ryle evolved into a major feud. The clashes between Hoyle and Ryle, and also between Gold and Ryle, are described in Mitton (2005).

45. *I'm going to Paris to a symposium*: From 30 July to 6 August 1958 Hoyle attended a symposium in Paris organized by the International Astronomical Union. See Hoyle (1959). He subsequently went to Moscow for another meeting of the International Astronomical Union and then returned to Caltech and eventually to Cambridge, England.

46. *synthesis of elements in various types of stars*: Together with the American nuclear physicist William Fowler and the two British astrophysicists Margaret and Geoffrey Burbidge, Hoyle developed an ambitious theory of element formation that gave a satisfactory explanation of the abundances of almost all the elements and their isotopes. The theory, colloquially known as the B^2HF theory, was published in 1957 and quickly became recognized as a classic, the foundation for further work in nuclear astrophysics. Since the theory made no use of a hot and compact past of the universe, it was widely seen as a problem for the primordial alternative favoured by Gamow and his collaborators.

47. *cosmological version of Lyell's uniformitarian geology*: In *Principles of Geology*, his classic three-volume work published between 1830 and 1833, the Englishman Charles Lyell argued that the science of geology must be based on the principle of uniformitarianism. According to this principle, the forces that have shaped the Earth in the past must be the same as those that can still be observed. The analogy between steady-state cosmology and Lyellian geology was discussed in a paper by the philosopher Stephen Toulmin (1962).

48. *nonetheless a triumph of the steady-state*: In a postscript of 1963 to an essay originally published in 1957, Bondi dealt with the B^2HF theory, which he considered "a tremendous triumph" of the steady-state theory. "This theory of the origin of heavy elements means that the steady-state theory has effectively passed a severe test," he wrote (Bondi 1966, p. 400).

49. *Oppenheimer admitted as much*: The famous American physicist J. Robert Oppenheimer was on the scientific committee of the 1958 Solvay congress. He commented on Hoyle's Solvay lecture that although the steady-state theory was, in his view, "quite wrong," it had provided the incentive for the B^2HF theory and in this way led to great progress in the understanding of the universe.

50. *corresponded to a very weak background radiation*: In 1940 the Canadian astronomer Andrew McKellar found that the excitation temperature of CN radicals in interstellar space was 2.3 K. Aware of this result, Hoyle (1949b) related it to Lemaître's cosmological model but not to the prediction of Alpher and Herman. In 1949 and on some later occasions he referred to McKellar's result as if it indicated a temperature in space of 1 K or less, rather than the higher temperature 2.3 K. With hindsight, the excitation temperature is a result of the CN radicals being bathed in the cosmic background radiation, but this was only recognized in 1966. See also Hoyle (1981) and Chapters 10 and 13.

51. *astronomers quickly plagiarized his discovery*: "Baade's announcement was made orally," Hoyle (1994, p. 263) recalled. "Surprising as it may seem, attempts were made to rob him of his priority over the age of the Universe, and, at the pinch, it was my minute that saved the situation."

52. *Baade was not pleased*: On 5 January 1953 the *New York Times* carried an article reporting that Shapley had doubled the scale of the universe. The same year the recently retired Shapley published his "discovery" in the *Proceedings of the National Academy of Science*, where it would not be refereed. A furious Baade characterized Shapley in letters as a "wind bag" and a "carnival barker" whose behaviour was "simply shameless." See Osterbrock (2001, p. 171).

53. *no respect at all for theoretical cosmology*: According to Baade's biographer (Osterbrock 2001, p. 205), he considered cosmology "a waste of time." Characteristically, when he spoke on the cosmic distance scale at a "Symposium on Cosmology" organized by the Australian National University in April 1959, he did not relate the subject to cosmology.

54. *he simply dislikes the cosmological principle*: At the 1957 Vatican symposium Baade expressed his disrespect for the cosmological principle on which most cosmological models relied. Heckmann (1976, p. 57), who was present at the symposium, says that Baade was convinced that the principle was "an admission to the thought-style of the time, to the socialist-communist tendency to see all humans as equal."

55. *the ages of the oldest stars*: In a lecture of 1960, Hoyle made it clear that he still considered the time-scale a problem: "The ages of the oldest stars in our own galaxy appear to be at least 1.5×10^{10} years, and may indeed be as high as 2×10^{10} years. If we accept the present-day estimates of the galaxies, . . . only Lemaître's cosmology with $\lambda \neq 0$, and the steady-state theory, survive the test." Hoyle (1961, p. 6). The symbol λ denotes the cosmological constant.

56. *Popper, who also turned up in your discussion with Whitrow*: In 1934 the Austrian-British philosopher Karl R. Popper published his main work *Logik der Forschung*, which in 1959 was translated into *The Logic of Scientific Discovery*. In 1949, after

positions in Austria and New Zealand, he was appointed professor at the University of London. In his public debate with Whitrow, Bondi defended Popper's criteria of science and Whitrow referred to "The important role of disproof in science, which has been so cogently argued by K. R. Popper." (Whitrow and Bondi 1954, p. 280). For details on Popper and cosmology, see Kragh (2013f).

57. *speaks the language of a physicist*: In a glowing review of *The Logic of Scientific Discovery*, written with the mathematical physicist Clive Kilmister, Bondi wrote that, "Popper speaks as a working scientist to the working scientist in a language that time and again comes straight out of one's heart." (Bondi and Kilmister 1959–1960).

58. *Einstein was quite enthusiastic*: In a letter to Popper of 1935, Einstein wrote: "Your book has pleased me very much in many ways: rejection of the 'inductive method' from an epistemological standpoint. Also falsifiability as the crucial element of a theory." (quoted in Kragh 2013f, p. 328).

59. *its foundation in the perfect cosmological principle*: Popper recalled that he had discussions with Bondi, some of them about the cosmological controversy. "I liked his [Bondi's] theory," he wrote in 1994, "but not the so-called 'cosmological principle' and even less its (temporal) extension." See Kragh (2012d).

60. *he really is a great philosopher*: Bondi continued praising Popper's philosophy of science many years after he had left cosmology. In his obituary of Popper in *Nature* Bondi wrote: "Although many scientists have little interest in the philosophy of science, . . . to me his thoughts came as a flash of brilliant light." (Bondi 1994).

61. *some advertising of his ideas*: "The essential characteristic of a scientific work, as Popper said, is this method of empirical disproof of a theory by experiments the outcome of which it has predicted incorrectly. . . . The theory, if it is to be a scientifically useful theory, must positively stick out its neck in order to run the risk of being disproved." (Bondi 1958–1959, p. 60).

62. *if he knew about it at all*: In a letter of 1952 Einstein commented on "the cosmological speculations of Mr Hoyle," which is his view were "too poorly grounded to be taken seriously." (quoted in O'Raifeartaigh et al. 2014). In fact, more than twenty years earlier Einstein had considered a steady-state model of the universe which anticipated the main features of the theory of Hoyle, Bondi, and Gold. However, he found the model to be unsatisfactory and did not publish it. See O'Raifeartaigh et al. (2014) and Castelvecchi (2014).

63. *Cosmology is practically non-existent in the Soviet Union*: During the Stalin and early post-Stalin era cosmological models of the kind discussed in the West were considered with suspicion as they were thought to disagree with the doctrines of dialectical materialism (Kragh 2013e). According

to these doctrines, the universe had to be infinite in both space and time, and creation was anathema. The steady-state theory was either ignored or criticized for resting on the hypothesis of continual creation of matter. Hoyle recalled: "Judge my astonishment on my first visit to the Soviet Union when I was told in all seriousness by Russian scientists that my ideas would have been more acceptable in Russia if a different form of words had been used. The words 'origin' or 'matter-forming' would be O.K., but creation in the Soviet Union was definitely out." (quoted in Kragh 1996, p. 263).

12

Paul Dirac and the Magic of Large Numbers

Interview conducted on 8 September 1963 at Dirac's house, 7 Cavendish Avenue, Cambridge, England. Language: English.

I had not originally planned to interview Dirac, who was after all a famous quantum physicist and not a cosmologist. But when I talked with Gamow in 1956 he told me about Dirac's ideas on large numbers and the universe, suggesting that I tried to arrange an interview. A few years later Hoyle made the same suggestion. They both told me that he was a man of few words with whom it might be difficult to keep a conversation going, but that it was possible. The interview went through only because I happened to know an American lady who had become a good friend of Mrs Dirac and regularly corresponded with her. At any rate, I finally came to meet the British mathematician and quantum genius in Cambridge (Figure 13). My health was not the best, but I decided to go to England anyway, realizing that an interview with Dirac was a rare thing and that at my age I would not get a second chance. He was surprised to learn that I wanted to hear about his cosmological theory rather than his work on quantum mechanics. I got the impression that he was pleased to speak of his more current ideas about the universe. Although it was rather difficult to engage him in a dialogue, under the circumstances I think the interview went fairly well.

While preparing for my conversation with Dirac I learned that he had recently been the subject of a detailed interview conducted by an American historian, Thomas Kuhn. I knew his name from a fine book he had written about Copernicus,[1] so I phoned him to find out more about the interview. It turned out that it had dealt only with Dirac's youth and early contributions to quantum theory, which was a relief to me. When I told him about my own more amateurish project, suggesting that cosmology was about to enter a phase no less revolutionary than the one associated with Copernicus and his time, he did not seem particularly interested.

Figure 13 Paul A. M. Dirac.

CCN Professor Dirac, as we agreed, our conversation will be restricted to cosmology, and I promise to keep it brief. But first, you are Lucasian Professor here in Cambridge, a chair first occupied by Newton and later by other famous physicists. Do you feel yourself to be part of Newton's heritage, a kind of twentieth-century Newton² perhaps?

PD No.

CCN Well . . . you are a renowned quantum theorist and most people will be surprised to learn that you have done interesting work in cosmology, a science that is after all quite different from quantum mechanics. It will be useful to go back in time, I think, back to the period before the war. Did you follow the developments in cosmology at the time, the expanding universe for example?

PD I knew about it.

CCN And how did you know about it, was it by reading the scientific literature or perhaps by listening to colleagues in physics and astronomy?

PD It was mostly by listening. I read some of the literature, but not systematically. I prefer listening to reading. As far as I recall I was introduced to the subject by Robertson,[3] who was an expert in cosmology and general relativity but also had an interest in quantum theory. And then Lemaître gave a talk in Cambridge.[4] I don't remember when it was, but it was shortly after he had proposed the idea of a primeval atom. I found his talk interesting and afterwards we went for a walk, when he told me more about it. It was somewhat controversial at the time, I think, but I had no problem with an expanding universe born a definite time ago.

CCN When you wrote your letter to *Nature*[5] in 1937 you didn't refer to Lemaître, at least not explicitly, but you referred to Eddington and Milne, so I suppose that they played a role as catalysts in your thinking about the universe, that they were sources of inspiration?

PD …

CCN Hmm, perhaps I should rephrase my question: did Eddington and Milne act as catalysts in turning your interest toward cosmology?

PD In a sense, at least some of their ideas did. They had both been my teachers[6] when I came to Cambridge and I knew them rather well. Milne had the idea of two different time-scales and Eddington tried to explain the relationship between the constants of nature by means of his new theory, claiming that it was an extension of quantum mechanics.

CCN The way you express yourself suggests that you did not agree with Eddington. Is that right?

PD Yes.

CCN And why did you not agree with him?

PD He used quantum mechanics incorrectly.[7] I don't think he really understood the theory, but his attempts to explain the fine-structure constant and the proton-to-electron mass ratio were interesting nonetheless. These numbers are small and I thought that the large dimensionless numbers needed a different explanation.

CCN Can you explain the essence of the theory you proposed in 1937 and elaborated in your article the following year?

PD Yes.

CCN Hmm. Can you please elaborate? I would be grateful for more information.

PD I see. It's very simple. It was known at the time, from Eddington and Weyl in particular, that the ratio of the electrical over the gravitational force between a proton and an electron is a very large number. It is about 10^{39}. It is sometimes called Weyl's number,[8] but I have not looked up where he mentions it. Then I noticed that if one measures the age of the universe in atomic units of time,[9] as given by a certain combination of constants of nature, another very large number of the same order turns up. Now, this could just be a coincidence, but in my view it is a significant relationship implying that the gravitational constant decreases slowly with cosmic time. For a billion years ago gravity must have been considerably stronger than it is today. It follows from the equations. One may also say that Weyl's number is so large because our universe is so old. It reflects a particular historical epoch, and in this sense one might call it accidental or, what may be a better term, contingent.

CCN I understand your reasoning, but it seems to me that your conclusion is based on assumptions that are somewhat arbitrary.[10] For example, if one used a different unit of time the age of the universe would no longer be given by the number 10^{39}. And you are also assuming, at least implicitly, that the other constants, such as the mass and charge of the electron, are unaffected by the epoch of time. Or is there something I have misunderstood?

PD No, you seem to have understood it. But although your objections are correct, they are not of great weight. The important thing is that the assumptions I make are reasonable and natural. We can safely assume that the atomic constants, such as the elementary charge, do not vary in time. The atomic time unit of approximately 10^{-23} seconds is the only natural one; it corresponds to the time it takes for a photon to pass the diameter of a classical electron.[11]

CCN I see . . . a natural choice, hmm. But one is not forced to draw the conclusion, is that right?

PD Yes, you are right, it is a hypothesis. In 1937 I also pointed out that the number of particles in the universe—what Eddington called the cosmical number—is close to the square of 10^{39} and that it presumably means that the number of particles increases quadratically with the cosmic period.[12] However, ordinary quantum mechanics does not allow the spontaneous generation of real particles, so the following year I retracted the hypothesis. More recently I have had second thoughts about this matter, contemplating that perhaps the hypothesis wasn't so crazy after all—or perhaps there is something to it precisely because it is so crazy. I have not made up my mind yet.

CCN Excuse me for being so critical, but I'm merely curious. We have two very large numbers turning up in nature and then you say that they are connected in some significant way. Isn't it rather vague? How close do the numbers have to be?[13] I mean . . . in your paper of 1938 you say that the mean density of the universe is about the same as the Hubble constant, if expressed in appropriate units, and yet the numbers differ by an enormous factor, of no less than 100,000. Can we really say that two numbers differing that much are roughly equal? Some would say that it's stretching the meaning of "equal" a bit too far.

PD I think they can be said to be roughly equal. To me 10^{-39} and 10^{-44} are of the same order of magnitude.

CCN Thank you. I wonder if you were the first to come up with the hypothesis of a gravitational constant varying in time. Are you aware of any predecessors?

PD No, I was aware of Milne's theory and he had some ideas, but they were different from mine. There are some people who have suggested that other constants might vary,[14] such as the fine-structure constant. I think they are wrong, but it's a possibility and one should not rule it out.

CCN The theory you proposed in a more mature form in 1938 predicts a definite change in the gravitational constant from which you derived a cosmological model that has features in common with Lemaître's model, as far as I can see. It starts in a singular state. Did you think at the time that the decrease in gravity could be subjected to experimental tests? And what about today, with the progress in precision measurements?

PD You are asking two questions. I cannot answer two questions at the same time.

CCN Oh, sorry, let me try again. . . . No, first I would like to know how people responded to your hypothesis. From what I have read it was considered controversial, at least in some circles, and you were attacked by Herbert Dingle who even called it . . . let me see, yes, he coined the word "cosmythology"[15] for the kind of theory you proposed. I like this word, I must admit. Did the criticism worry you?

PD No.

CCN Could you be more specific?

PD No.

CCN Well, it's interesting, because many years ago I had a conversation with Eddington and asked him the same question. As you know, he was also attacked by Dingle and he responded in about the same way as you—he didn't take him seriously. Do you recall other responses to the theory, positive or negative?

PD It's many years ago, but I remember that Chandrasekhar liked it[16] and that he thought it might be of value in understanding stellar evolution. Bohr did not like it.[17] Gamow was in Copenhagen at the time, and he reported Bohr's attitude to me. Perhaps he found it too speculative. But I thought the idea was a nice one and paid little attention to what other people thought of it.

CCN Let me now return to the previous question, or let me split it up into two. First: is it correct that your cosmological model has features in common with Lemaître's?

PD Not really, only in the sense that the universe is expanding and has a beginning in time. But there is no room for a cosmological constant in my model, whereas it is essential to Lemaître's model, and whereas his universe is closed and finite, mine is flat and infinite. It results in a definite recession law for the galaxies and therefore also in a definite age of the universe,[18] but unfortunately one that is very low and apparently in contradiction to the age of the Earth. This is a problem, but not necessarily a fatal one. I think it is solvable.

With regard to Lemaître I may mention that he had this idea that the quantum uncertainties in the early universe provide an escape from the difficulty of a universe evolving deterministically. It is a difficulty because, if this were the case, we cannot explain the complexity of our universe by starting in a very simple and uniform state. I think it was a clever idea and in a lecture I gave in Edinburgh[19] shortly before the war I presented it in a more elaborate form. I also speculated that the laws of nature may depend not only on the epoch but also on the position in space,[20] but it remained a speculation. I now realize that I interpreted the space–time symmetry of relativity theory too literally.

CCN Your entire cosmological theory is based on a principle in which I understand you have a great deal of faith and which in your paper of 1938 you called the "fundamental principle." It seems to be of a very general nature. Can you give a brief formulation of it, please?

PD I referred to the principle that all the very large dimensionless numbers occurring in nature are simple powers of the epoch, with coefficients of the order unity. I now prefer to think of it as the large-number hypothesis[21] because it is concerned with very large combinations of constants only.

CCN Right, so it's not only the number 10^{39} but also even larger numbers such as 10^{78} and perhaps, let me see . . . 10^{127}.

PD You mean 10^{117}, I suppose, but I have not as yet found any use for this number. They should be understood as rough orders of magnitude, but that is precise enough. It is a fascinating thought that perhaps all events correspond to properties of a single large number.[22] It may signify some deep connection between pure mathematics and physics that remains to be explored.

CCN I have myself played around with some of these numbers[23] and come up with one of the magnitude 10^{117} or 10^{120} that may have some significance. If one transforms the mass of the universe into energy by means of Einstein's mass–energy relation and multiplies it by the age of the universe, the result actually becomes of that order. Do you think this is worth consideration?

PD No.

CCN Too bad. Now, to return to my earlier question: Can the varying-gravity hypothesis be tested by means of experiment?

PD Yes, but the decrease in gravity is so slow[24] that it cannot be measured directly, at least not by present technology, I think. With the present value of the Hubble constant the annual decrease is only about 10^{-10}, which is not practically measurable. But there are other more indirect ways to test the prediction, for it has some astrophysical and geophysical consequences.

CCN Yes, and I would like to return to these consequences, but first I'm curious to know how you feel about the current situation in cosmology. For the last decade or so we have witnessed an epic battle between the evolutionary relativistic universe and the steady-state theory, with the latter theory now being on the retreat as far as I can tell. What's your opinion of the steady-state theory of the universe?

PD I have formed no opinion of it except that it is probably wrong. It includes some appealing features, such as the continual creation of matter, but it is irreconcilable with my large-number hypothesis[25] and a gravitational constant varying in time. I don't believe in it.

CCN I had a conversation with Hoyle and Bondi a few years ago, but forgot to ask them about your ideas. Did you know of the steady-state theory early on?

PD Yes, I once served as Hoyle's supervisor,[26] but at the time he worked in quantum electrodynamics and not in cosmology. Later I also had Dennis Sciama as my student,[27] he was interested in the new theory of the steady-state universe. I think I first heard about it at a seminar Hoyle gave at the Cavendish Laboratory[28] in 1948, it must have been before the papers were published. Heisenberg was visiting Cambridge at the time and I remember he was present at Hoyle's seminar. I don't think he convinced any of us.

CCN Excellent, now that is clarified. When I interviewed Gamow some years ago in Colorado he told me about the possibility of a weak background radiation that originated in the very early universe. It hasn't been detected, as far as I know, but Gamow thought it was real and that it might prove his idea of a hot beginning of the universe. Is that something you have heard of or have considered?

PD I only know it from Gamow, but as far as I can see it has no place in my cosmology,[29] which I told him. It is supposed to be due to a fundamental decoupling between matter and radiation at a certain time that cannot be brought into agreement with the large-number hypothesis.

CCN It is not obvious to me why . . . no, it doesn't matter. For a period of more than twenty years you stopped working on cosmology and then, two years ago, you made a brief comment in *Nature* on the subject. Can I take your long silence as an indication that you were dissatisfied with your theory?

PD I was not dissatisfied, not really, but I could not see how to improve the theory and how to avoid the much too short age of the universe, for example. I did not publish on cosmology simply because I had no new ideas to communicate. If one has nothing new to say, one should stay silent. My mother taught me that.

CCN Your mother?

PD Yes.

CCN Sorry, I didn't mean to interrupt you. Please continue.

PD Then, after the war, Teller came up with his critique which was widely believed to refute my theory. I did not think so, but most people did, and I guess it dissuaded me from taking another look at it. Jordan tried to revive it[30] and he wrote me letters about his ideas, but I did not find his theory convincing and decided to continue with my work on improving the equations of quantum mechanics and its whole foundation.

CCN And the critique you refer to was Teller's counterfactual argument[31] based on a scenario of what the Earth would have looked like if the gravitational constant had been greater in the past. He showed from various relations in solar physics that your theory would lead to the consequence that in the Cambrian era the seas would have been boiling with no living creatures in them. Since palaeontologists assure us that there was in fact a richness of life in the Cambrian seas, such as trilobites, worms, molluscs, and the like, he inferred that the hypothesis must be wrong. Is that a fair summary of Teller's argument?

PD Yes.

CCN Did you find his argument compelling?

PD No, but other people did. Gamow liked my idea very much, but he nonetheless accepted Teller's argument as a disproof. In fact, I think he was himself somehow involved in it, because he told me he had discussed the question with Teller.[32] I did not respond to the paper because my theory also faced the problem of the short time-scale. The Hubble time was thought to be less than two billion years, but we now know that it is much longer, perhaps as long as twenty billion years. Not only did the extension remove the time-scale difficulty, it also weakened Teller's argument[33] because it depended on the age of the universe. I no longer see any reason why my theory should not be correct. I told Gamow about it,[34] but without convincing him.

CCN When was that?

PD Three or four years ago. Gamow recently sent me a popular book[35] in which he discusses the question. I think he still considers it an open question whether or not the gravitational constant varies in time. He doesn't want to commit himself.

CCN Right, it appears that Teller's objections have not deterred physicists from taking up your theory. There has recently been a good deal of interest in it, not only as a cosmological theory but also from the point of view of geophysics. Are you following this revival of interest in your ideas?

PD Only to a limited extent. I am not really interested in geophysics, but I know about some of the work done by Jordan and Dicke. There was a recent paper in which Jordan said that the modern picture of the crust of the Earth amounts to a proof of my idea of a varying gravitational constant.[36] Perhaps he exaggerates.

CCN And Robert Dicke has also dealt with the large numbers in several publications investigating the astrophysical and geophysical implications of your hypothesis. But it appears that his ideas are rather different from yours, in so far as he does not consider the age of the universe to be given by the large-number combination of constants involving the gravitational constant. He says something to the effect that the present epoch is conditioned by our existence[37] . . . It sounds strange to me, I must admit, so I wonder if I have understood what he means. Could you perhaps explain what it's about?

PD Yes, if you want. About two years ago Dicke wrote a paper in *Nature*[38] and the editor requested me to comment on it. It is correct that Dicke argues that some of the large-number coincidences, as I think he calls them, come about as the result of our existence as an advanced biological species based on carbon molecules. He offers an alternative explanation of the large-number relationship that, according to my view, is an argument for the force of gravity decreasing with cosmic time.

CCN I see, and in this comment of yours, did you criticize his suggestion of a connection between human observers and the age of the universe?

PD No, I merely pointed out that the alternatives differ with regard to life in the future. Dicke's view is pessimistic, mine is not.[39] The question can only be decided by means of measurements, which can be done in principle, but as I said, the accuracy of current measurements is insufficient. For the moment it is largely a matter of taste, and I have not concealed what my taste is.

CCN Now, this was two years ago, and I assume . . . I mean, are you busy with developing your theory[40] and perhaps turning it into a more easily testable version?

PD No.

CCN But will you do it at some day?

PD Perhaps.

CCN I have one last issue that I would like your opinion of, and that relates to your famous prediction of antiparticles[41] which you proposed in the early 1930s on the basis of your new wave equation of the electron. We know that these antiparticles exist and according to some physicists they even play a cosmological role. Is that something you have thought about?

PD I mentioned it in the lecture I gave when receiving the Nobel Prize[42] in Stockholm.

CCN But that was thirty years ago. I mean, have you thought of it more recently? Does the idea of cosmic antimatter fit into your present ideas of a universe governed by a gravitational constant varying in time?

PD I think not.

CCN There has been some discussion recently about possible "antiworlds." I remember a short paper by an American physicist who suggested that there might be an entire universe made up of antimatter[43]—separated from ours, of course. I can't remember his name, but it's a fascinating thought, wouldn't you say?

PD No.

CCN Well, I also had in mind some ideas of a somewhat less speculative nature proposed quite recently by Oskar Klein and Hannes Alfvén in Sweden.[44] Are you familiar with these ideas?

PD No, but I have known Klein for a long time. I first met him in Copenhagen. He used to work in quantum field theory.

CCN Hmm, let me try another angle. I understand that according to your quantum theory particles and antiparticles are completely symmetric. The only reason that we call the electron a particle and the positron an antiparticle is that the latter kind of particle is so rare, and the same goes for the proton and the antiproton. And yet antimatter seems to be almost completely missing[45] in our universe. Is there any reason for this asymmetry?

PD There must be a reason, but I cannot see what it is.

CCN Aha, so this might be a problem in cosmology and possibly one that can only be explained by the conditions of the very early universe?

PD . . .

CCN Hmm, let me try again. Professor Dirac, do you think that the asymmetry between matter and antimatter counts as a cosmological problem that ought to be addressed?

PD Possibly.

CCN Sorry to be so persistent, but there is a related question I would like to ask. In the same paper of 1931 in which you predicted the positron you also came up with the suggestion of magnetic monopoles.[46] As far as I know, these particles have not been discovered, but nonetheless they may exist. It's possible that they are just very rare. Do you think monopoles may play a role in cosmology[47]?

PD I doubt it, although there were some speculations in the 1930s that they might be present in cosmic rays. The theory of the magnetic monopole is very satisfactory, but it is quite possible that the particles do not exist in nature. In a sense they ought to exist, but for some reason nature may have chosen not to make use of them.[48] What that reason for this is, I don't know. Perhaps they existed in the very early universe, but that is just a speculation.

CCN Thank you for your answer. Well, let's call it a day. It has been a great experience listening to the words of one of the world's most famous physicists. I really appreciate it. Please give my best regards to your wife.

Notes

Paul Adrien Maurice Dirac was born on 8 August 1902 in Bristol, England and died on 20 October 1984 in Tallahassee, Florida. Dirac's father was a Swiss national, and Paul was registered as Swiss by birth, only acquiring British nationality in 1919. Paul Dirac studied at the University of Bristol, graduating in 1921 as an electrical engineer, but with no employment. In 1923 he became a research student at St John's College, Cambridge, where he quickly transformed himself into a promising physicist. Immediately after Werner

Heisenberg's proposal of quantum mechanics in August 1926, he developed his own version of the theory in the form of a more general, algebraic formulation. He applied the theory to a number of problems in quantum statistics (Fermi–Dirac statistics) and radiation theory. In 1927 he formulated the basic equations of quantum electrodynamics and the following year, at the age of 25, he presented a new relativistic wave equation for the electron, the Dirac equation. This very important work was recognized as revolutionary and led to extraordinary results. Not only did it explain the electron's spin and the details of the hydrogen spectrum, it also led to several new predictions. Analysing the equation, in 1931 Dirac suggested the existence of positively charged "antielectrons," soon to be identified with the positrons found in the cosmic radiation. He also predicted the existence of magnetic monopoles, but these particles have escaped detection.

Apart from frequent travels, Dirac stayed in Cambridge until he retired in 1969 (see Figure 13). In 1930 he was elected a fellow of the Royal Society and in 1932 Lucasian Professor of Mathematics. The following year he received, jointly with Erwin Schrödinger, the Nobel Prize in Physics. He deeply influenced a generation of physicists with the classic textbook *The Principles of Quantum Mechanics* first published in 1930 and subsequently in several revised editions. His most important scientific contributions, all of them in quantum theory, dated from the period 1925–35, after which he gradually turned away from mainstream physics. One example of his growing heterodoxy was a cosmological theory based on a varying gravitational constant that he proposed in 1937–8 and returned to only in the 1970s. Another example was a theory of the classical electron. He continued working on an improved theory of quantum electrodynamics, but his efforts met with little success. In the 1960s Dirac also worked extensively on the general theory of relativity. In 1971 he took up a professorship at Florida State University, Tallahassee, where he stayed for the rest of his life. During this period he devoted much of his time to his own version of cosmological theory.

Biographical sources: Kragh (1990); Farmelo (2009); Dirac (1977). A complete bibliography is included in Kragh (1990). See also the AIP interviews of 1962–3 (see AIP Interviews).

1. *a fine book he had written about Copernicus*: In 1957 Thomas S. Kuhn, a historian and philosopher of science, published *The Copernican Revolution*. His most important work, *The Structure of Scientific Revolutions*, appeared five years later, but at the time of the telephone conversation CCN was not aware of it.

2. *a kind of twentieth-century Newton*: Dirac was appointed Lucasian Professor of Mathematics in 1932, succeeding Joseph Larmor. He was often compared with Newton, who held the chair between 1669 and 1702, but to Dirac neither the chair nor his illustrious predecessors meant much. CCN was

slightly misinformed, as the first chairholder was not Newton, but Isaac Barrow (1663–9).

3. *I was introduced to the subject by Robertson*: A reference to the American physicist Howard P. Robertson, whom Dirac had first met in Göttingen in 1927. Dirac spent the autumn term of 1931 at Princeton University, where Robertson was professor. During his stay they got to know each other well. Most likely they discussed cosmology, a science which at the time was experiencing dramatic changes. Dirac returned to Princeton in 1934, this time to the new Institute for Advanced Study, where Einstein had settled. For Robertson, see also Chapters 5, 6, and 7.

4. *Lemaître gave a talk in Cambridge*: On 25 April 1933 Lemaître gave a talk on "The Primaeval Universe" to the Kapitza Club, an informal discussion and lecture club for Cambridge physicists. Dirac was elected a member of the Kapitza Club in the autumn of 1924 and remained an active member for many years. See Kragh (1990, p. 224), on which part of this chapter is based.

5. *When you wrote your letter to Nature*: This was Dirac (1937), which was based on the view of "current cosmological theories, [according to which] the universe had a beginning about 2×10^9 years ago, when all the spiral nebulæ were shot out from a small region of space, or perhaps from a point."

6. *They had both been my teachers*: For a couple of months in 1925 Milne acted as Dirac's supervisor, substituting for Ralph Fowler who was on leave in Copenhagen. While a student Dirac also followed Eddington's lectures on the theory of special and general relativity. With Eddington's help he wrote one of his first scientific papers on this subject.

7. *He used quantum mechanics incorrectly*: In a letter of 21 November 1935 to the Danish physicist Christian Møller, Chandrasekhar wrote from Cambridge: "Dirac thinks that Eddington is mad—so do all of us!" (quoted in Kragh 1991, p. 111).

8. *Weyl's number*: The German mathematician Hermann Weyl discussed the number 10^{39} in a paper of 1919 and the following year it was taken up by Eddington. It can be found in the physics literature even earlier, first in Zöllner's 1872 book on comets mentioned in Chapters 2 and 3.

9. *atomic units of time*: Dirac took the time unit to be $e^2/mc^3 = 10^{-23}$ s, where e and m refer to the electron's charge and mass, respectively. This unit of time was sometimes referred to as a "chronon" and thought to be a minimum period for physical processes. As Dirac was aware, there are other ways of constructing atomic time units, for example h/mc^2 and h/Mc^2, where M is the mass of a proton.

10. *assumptions that are somewhat arbitrary*: For a critical evaluation of Dirac's reasoning and his liberal use of the concept of order of magnitude, see Klee (2002). The relation $e^2/GmM \sim t$ only implies $G \sim t$ if is assumed that e, m, and M are independent of time. Dirac claimed that this was the case, but

without providing arguments for the claim. Indeed, in the 1960s Gamow suggested from the very same relation that $G =$ constant and $e^2 \sim t$. There have also been proposals that the masses of elementary particles vary in cosmic time.

11. *diameter of a classical electron*: The so-called classical electron has a diameter of the order of e^2/mc^2, meaning that a photon will pass it in a time given by $e^2/mc^3 = 10^{-23}$ s.

12. *the number of particles increases quadratically with the cosmic period*: Dirac (1937) gave an estimate of 10^{78} for the number of particles in the visible universe, which, he noticed, is the square of its age t in atomic time units. Therefore, "we see that the number of protons and neutrons in the universe must be increasing proportionally to t^2." Although he abandoned the idea in his theory of 1938, it reappeared in some of his later versions of cosmology.

13. *How close do the numbers have to be?*: Dirac took Hubble's constant in atomic units to be 10^{-39} and the mean density of the universe, in the same units, to be 10^{-44}. He concluded that "the average density of matter is of the same order of smallness as Hubble's constant." (Dirac 1938, p. 203). In a later paper Dirac was even willing to consider the number 10^{28} as belonging to the large-number cluster 10^{39}. He found the discrepancy between the two numbers "not significant in view of the rough nature of the LNH [large-number hypothesis]." See Kragh (1990, p. 245).

14. *that other constants might vary*: In the mid-1930s there were a few speculations that Planck's quantum constant, and therefore the fine structure constant, might vary in time. The hypothesis was used to explain the galactic redshifts without accepting a universe in expansion. As mentioned in Chapter 8, there were also proposals that the velocity of light varies in time.

15. *he coined the word "cosmythology"*: In his polemical article on "Modern Aristotelianism," Dingle (1937) attacked the theories of Milne, Eddington, and Dirac. "Instead of the induction of principles from phenomena," he wrote, "we are given a pseudo-science of invertebrate cosmythology." He regretted that Dirac was the latest "victim of the great Universe mania." See also Chapter 8.

16. *Chandrasekhar liked it*: Chandrasekhar (1937), who in a letter to Dirac of March 1937 told him that he was "quite excited" about the new hypothesis of a varying gravitational constant (Kragh 1990, p. 233). His excitement soon evaporated.

17. *Bohr did not like it*: Gamow (1967, p. 767) recalled: "The first criticism of this idea was made by Bohr. I still remember him coming to my room (I was visiting Copenhagen at the time), with the fresh issue of *Nature* in his hands, saying: 'Look what happens to people when they get married.'" Paul Dirac married Margit Balazs, a widowed sister of the physicist Eugene Wigner, on 2 January 1937.

18. *a definite age of the universe*: For the radius or scale factor R of the universe, Dirac deduced that it varied as $t^{1/3}$, which implies that the age of the universe relates to the Hubble time T as $T/3$. For comparison, the Einstein–de Sitter model predicts $t^{2/3}$ and $2T/3$. With the accepted value of the Hubble constant it meant an age for the universe of only 700 million years, or about one-third of the age of the Earth as known at the time.

19. *a lecture I gave in Edinburgh*: On 6 February 1939 Dirac was awarded the James Scott Prize by the Royal Society of Edinburgh. He delivered a lecture on "The Relation Between Mathematics and Physics" in which he argued for an intimate connection between fundamental physics and what he called beautiful mathematics. Dirac's indebtedness to Lemaître's primeval atom hypothesis is particularly clear in this lecture. See Dirac (1939) and Kragh (2011, pp. 175–177).

20. *but also on the position in space*: Dirac (1939, p. 128) said that one should expect the laws of nature "also to depend on position in space, in order to preserve the beautiful idea of the theory of relativity that there is fundamental similarity between space and time." He did not follow up on this very radical suggestion. As mentioned in Chapter 4, in the early part of the century Seeliger contemplated the possibility that the law of gravitation might vary in space.

21. *the large-number hypothesis*: "We may take it as a general principle," Dirac wrote in his paper of 1937, "that all large numbers of the order 10^{39}, 10^{78} ... turning up in general physical theory are, apart from simple numerical coefficients, just equal to t, t^2 ... where t is the present epoch expressed in atomic units." This is known as Dirac's large-number hypothesis, a name that came into general usage in the 1970s and is often abbreviated LNH.

22. *all events correspond to properties of a single large number*: "Might it not be," Dirac (1939, p. 129) wrote, "that all present events correspond to properties of this large number, and, more generally, that the whole history of the universe corresponds to properties of the whole sequence of natural numbers?"

23. *played around with some of these numbers*: This idea of relating the mass–energy and age of the universe to the cube of Dirac's number was informally suggested by Stanislaw Ulam (1972, p. 274). It remained a speculation. A few other physicists, including Fermi, suggested dimensionless combinations of natural constants of the order of 10^{120}. See the survey of numerical coincidences in Zimmerman (1955). Incidentally, the ratio of the theoretically calculated cosmological constant (or dark energy density) and the one obtained from observation is of the same order. The huge discrepancy is known as the "cosmological constant problem."

24. *the decrease in gravity is so slow*: Dirac's theory implied a variation in the gravitational constant G given by the expression $\Delta G/\Delta t = -3HG$. With the value of Hubble's constant H accepted in 1938, the predicted annual decrease corresponded to about 10^{-10}, whereas it was closer to 10^{-11} using

the value known in the early 1960s. The Lunar Laser Ranging Experiment that started in the late 1960s measures the precise distance between the Earth and the Moon, from which a possible temporal variation in G can be inferred. Analysis of data using about forty years of observations has resulted in a variation of $\Delta G/G\Delta t = (2 \pm 7) \times 10^{-13}$ per year.

25. *it is irreconcilable with my large-number hypothesis*: For the steady-state theory, see Chapter 11. According to this theory the large-scale features of the universe are the same at any time, which rules out the large-number hypothesis in Dirac's sense. Nonetheless, the hypothesis played an important role in the formation of Hoyle's theory and it also featured prominently in the work of the leading steady-state protagonist William McCrea. He claimed that "the model conforms to Dirac's principle . . . [and] in so far as we accept this principle, we can take it as favourable to the steady-state hypothesis." See Hoyle (1982, p. 50) and McCrea (1950, p. 8).

26. *I once served as Hoyle's supervisor*: As a graduate student in Cambridge, Hoyle was supervised by Rudolf Peierls, then by Maurice Pryce, and, in 1938–9, by Dirac. See Hoyle (1994, pp. 133–134, 218).

27. *I also had Dennis Sciama as my student*: Sciama earned his PhD in 1953 under Dirac. From about 1955 to 1965 much of his work in cosmology, especially on the formation of galaxies, was done within the framework of the steady-state theory. He only abandoned the theory in 1966, primarily because of its inability to explain the distribution and redshifts of quasars. As a professor at Cambridge and later at Oxford, Sciama was an influential figure in the astrophysics and cosmology community.

28. *a seminar Hoyle gave in the Cavendish Laboratory*: Hoyle's seminar took place on 1 March 1948. The steady-state theory was announced in two papers that appeared in print in the autumn of 1948, one by Bondi and Gold and the other by Hoyle. Both papers referred to the large-number coincidences and Dirac's theory. However, in the steady-state theory the large numbers were permanent, in sharp contrast to Dirac's ideas.

29. *it has no place in my cosmology*: Many years after the discovery of the cosmic microwave background Dirac dismissed its standard explanation with the argument that the decoupling time, expressed in atomic units, is about 10^{26}. This number, he said, is incompatible with the large-number hypothesis (Kragh 1990, p. 244).

30. *Jordan tried to revive it*: The German theoretical physicist Pascual Jordan was an admirer of Dirac's hypothesis of a varying gravitational constant. He took up Dirac's cosmology as early as 1938 and after World War II he developed it in his own way in a series of books and articles, including *Schwerkraft und Weltall* [Gravitation and Universe] from 1952 and a review article (Jordan 1962). However, his theory was not highly regarded by the majority of astrophysicists and cosmologists. See Kragh (2004, pp. 175–185).

31. *Teller's counterfactual argument*: Teller argued that the surface temperature of the Earth would vary with the gravitational constant raised to the power of 9/4. From this he inferred that, according to Dirac's hypothesis, 200 or 300 million years ago, "We are led to expect a temperature near the boiling point of water." On the other hand, he realized that "our present discussion cannot disprove completely the suggestion of Dirac." (Teller 1948, p. 802).

32. *he had discussed the question with Teller*: Teller (1948) acknowledged discussions with Gamow and the astrophysicist Martin Schwarzschild—a son of Karl Schwarzschild—during the tenth Washington Conference on Theoretical Physics in 1947, the topic of which was "Gravitation and Electromagnetism."

33. *it also weakened Teller's argument*: According to Teller, the temperature in the past would depend on the ratio between the present age and the age in the past. With an age of the universe of 12 billion years the temperature of the Earth in the Cambrian era would reduce to about 45 °C and thus allow life on Earth.

34. *I told Gamow about it*: In a letter of 10 January 1961, Dirac wrote: "It was a difficulty with my varying gravitational constant that the time scale appeared too short, but I always believed the idea was essentially correct. Now that the difficulty is removed, of course I believe more than ever. The astronomers now put the age of the universe at about 12×10^9 years, and some even think that it may have to be increased to 20×10^9 years, so that gives us plenty of time." (quoted in Kragh 1990, p. 237). In reality, even the extended time-scale did not give "plenty of time." The values cited by Dirac were either Hubble times or ages of the universe on the assumption of an Einstein–de Sitter universe. In 1958 Sandage estimated 13×10^9 years as an upper limit for a universe of the Einstein–de Sitter type, which corresponds to 6.5×10^9 years for the Dirac model, a value smaller than the age of the oldest stars.

    In the same letter Dirac mentioned the possibility of a varying fine structure constant. In that case, "the chemistry of the early stages [of the Earth] would be quite different, and radio-activity would also be affected." He probably had in mind that the hypothesis might lead to an age of the Earth of less than 4.5×10^9 years. Gamow discussed modifications of Teller's argument on several occasions, arguing that it remained basically valid. See Kragh (1991) for his discussion with Dirac in the 1960s.

35. *Gamow recently sent me a popular book*: Gamow (1962, pp. 138–141) discussed the question of a varying gravitational constant. Although he concluded that Dirac's hypothesis was unlikely, he also thought that it could not be ruled out.

36. *a proof of my idea of a varying gravitational constant*: In a review paper of 1962, Jordan concluded: "I think that our present knowledge of the earth, . . . makes the correctness of Dirac's hypothesis an established fact, including also the

puzzling circumstance that during the formation of the Earth, κ must have been quite considerably greater than it is now." (Jordan 1962, p. 600). The quantity κ, sometimes known as Einstein's constant of gravitation, is equal to $8\pi G/c^2$.

37. *the present epoch is conditioned by our existence*: Dicke first pointed to what he called a "logical loophole" in Dirac's argument in 1958, when he noticed that it rested on the assumption that the epoch of humans is random. He suggested as an alternative that "The present epoch is conditioned by the fact that the biological conditions for the existence of man must be satisfied." (Dicke 1959, p. 33). In his paper of 1961 he likewise concluded that Dirac's large-number relations could be explained by "the existence of physicists now and the assumption of the validity of Mach's principle." (Dicke 1961, p. 441). Dicke's arguments make him a precursor of the anthropic principle, but in his papers from the period he did not suggest that the universe is somehow designed to be habitable for intelligent life. On Dicke's role in the emergence of modern big bang cosmology, see the interview with him in Chapter 13.

38. *Dicke wrote a paper in Nature*: Dicke (1961) is reprinted in Leslie (1990, pp. 121–124). It was followed by Dirac's untitled comment, his first announcement on cosmology since 1939.

39. *Dicke's view is pessimistic, mine is not*: Dirac wrote: "On this [Dicke's] assumption habitable planets could exist only for a limited period of time. With my assumption they could exist indefinitely in the future and life need never end. There is no decisive argument for deciding between these assumptions. I prefer the one that allows the possibility of endless life."

40. *busy with developing your theory*: Dirac was not busy at all. He only returned to his cosmological theory in 1973 and during the next eight years he developed it in various ways, some of which included spontaneous creation of new matter. See Kragh (1990, pp. 236–245). His many attempts to breathe new life into the theory failed. However, a minority of physicists closer to mainstream cosmology than Dirac continued to investigate the theory and construct more advanced versions of it without the weaknesses of the original theory. Maintaining the fundamental assumption of a varying gravitational constant, these theories claimed (and still claim) to account for both the microwave background and the primordial production of helium.

41. *your famous prediction of antiparticles*: In 1931 Dirac suggested the existence of positively charged antielectrons corresponding to solutions of the wave equation that he had found three years earlier. He had first thought that the antielectron was a proton, but now realized that it had to be a hypothetical particle with the same mass as the electron. The predicted particle was soon detected in the cosmic rays and then became known as a positron.

The negative antiproton was detected in accelerator experiments in 1955 and the neutral antineutron in 1956.

42. *when receiving the Nobel Prize*: At the end of his 1933 Nobel lecture Dirac speculated that there probably existed stars made up entirely by antimatter, that is, of positrons, antiprotons, and antineutrons. "In fact, there may be half the stars of each kind," he said. "The two kinds of stars would show exactly the same spectra, and there would be no way of distinguishing them by present astronomical methods." (quoted in Kragh 1990, p. 104).

43. *an entire universe made up of antimatter*: The American nuclear physicist Maurice Goldhaber pointed out that the Gamow–Alpher–Herman theory was based on the "unsatisfactory tacit assumption" of a starting scenario of nucleons with no antinucleons. He speculated that to preserve particle–antiparticle symmetry, "we have to consider models in which the cosmos and its possible counterpart, the 'anticosmos,' are somehow separated from the very beginning." (Goldhaber 1956).

44. *Oskar Klein and Hannes Alfvén in Sweden*: In 1963 Alfvén and Klein discussed the possibility that half of the visible universe consists of matter and the other half of antimatter, and they suggested a mechanism that would keep matter and antimatter separate. This idea was later elaborated by Alfvén into a plasma cosmology where the big bang was replaced by an original explosion of matter annihilating with antimatter. This kind of cosmology was for a time defended by a minority of physicists resisting the big bang theory. See Lerner (1991) and Peratt (1985).

45. *antimatter seems to be almost completely missing*: The problem of the missing antimatter only attracted wide attention in the 1970s, when physicists attempted to solve it by means of so-called grand unified theories. By then the problem was no longer to explain the missing antimatter, but to explain the result of the original (and hypothetical) annihilation of particles and antiparticles. According to theory, this primordial process would have led to a universe consisting almost only of photons. Consequently, why is there so much matter in the universe?

46. *suggestion of magnetic monopoles*: In 1931 Dirac proved that magnetic monopoles analogous to elementary electrical charges could be described consistently within the framework of quantum mechanics. The strength of a Dirac monopole would be about 68 times the strength of an electron (namely, half the inverse fine structure constant $h^2c/4\pi^2e^2 \cong 137$). The prediction attracted little interest and the few experimental searches failed to detect the particle. In 1948 Dirac improved the theory of the magnetic monopole, but in despite a few claims for its discovery it remains to this day hypothetical.

47. *monopoles may play a role in cosmology*: Only with the development in the 1970s of a new kind of magnetic monopole based on grand unified theory, primarily due to Gerardus t'Hooft and Alexander Polyakov, did the hypothetical

particles enter cosmology. They are believed to have been produced in copious numbers in the very early phases of the big bang. The lack of observed monopoles served as an important stimulus for the successful inflation theory of the early universe proposed in 1981 by Alan Guth. According to this theory, widely accepted today, the extreme scarcity of monopoles is explained by the primordial inflation of space.

48. *nature may have chosen not to make use of them*: Dirac tended to believe that since monopoles were consistent with the fundamental laws of physics, they would probably exist in nature. This kind of reasoning is an example of what in the history of ideas is known as the principle of plenitude. "Under these circumstances," Dirac wrote in 1931, "one would be surprised if Nature had made no use of it." (quoted in Kragh 1990, p. 220). Thirty years later he considered the existence of monopoles an open question that could only be decided by experiment. So far they have remained undetected.

13

Robert Dicke and the Big Bang

Interview conducted on 2 August 1965 at the Palmer Physical Laboratory, Princeton University. Language: English.

I had decided to end my series of private interviews after I completed my conversation with Dirac in 1963. It was time to put a stop to the game, if for no other reason than my advanced age. However, for decades I had made it my habit to follow the literature on cosmology, and I continued to do so, just following the law of inertia. When I learned from an article in the *New York Times* that the background radiation foreseen by Gamow and his collaborators had apparently been discovered, I realized that something very important was going on. I also realized that the pre-history of the radiation was apparently unknown—none of the press reports mentioned Gamow's name, nor Alpher's or Herman's.

I just couldn't resist the temptation of trying to arrange one last meeting, in this case with the senior physicist of the Princeton group involved in the discovery of the radiation. When I called Professor Dicke (Figure 14) and told him about my earlier interviews, he kindly agreed to meet me for a conversation in his office at Princeton University. He found it fascinating that I had started my series of interviews before he was born and also that I had talked with Einstein at a time when the expansion of the universe had not yet been recognized. I got the impression that he knew very little about the history of cosmology, even the more recent history, but I was not really surprised.

CCN Professor Dicke, let me first express my gratitude for your willingness to speak to an old man who's just an amateur in matters of cosmology. And also thank you for sending me the preprint of the paper that will appear shortly[1] in the *Astrophysical Journal* and to which we shall return. I have studied it carefully.

RD You're welcome.

Figure 14 Robert H. Dicke. Courtesy American Institute of Physics, Emilio Segré Visual Archives.

CCN Now, my impression is that you came relatively late to cosmology, following a somewhat irregular route that may perhaps be called non-astronomical. Could you please tell me how you came to work on the subject of the universe and its evolution in time?

RD Well, the short answer is gravity. My first scientific work of any value was in radar physics, more specifically on microwave radar; it was an outgrowth of what we did during the war at the Radiation Lab.[2] When I came to Princeton I thought of using my skills in this area to do radioastronomy, but the astronomers were simply not interested[3] and thus nothing came of it. I'm still surprised by their cool reception. Instead I focused on atomic physics, quantum optics and things like that—really all aspects of the interaction of matter and radiation as described by quantum mechanics. I wrote an introductory textbook on the subject which is still in use.

    Then in the mid-1950s I became seriously interested in gravitational physics, and of course gravity is intimately connected to cosmology, although for a time cosmology was rather peripheral to me, as it was to most other physicists in the field. I was more interested in how to test Einstein's theory of gravitation, to extend and refine the classical tests by means of modern experimental methods. By that time I had established a small group of physicists and students interested in general relativity. Wheeler was around,[4] he has been very helpful.

CCN So you and your group at Princeton were part of what some physicists refer to as the "renaissance of general relativity"?[5] I can't remember where I came across the phrase.

RD Oh yes, very much so, and we still are. It's a vigorous and exciting field, both when it comes to theory and experiment … Einstein would have been surprised. There are today several strong research groups in the area, whereas ten years ago there was hardly a single one. When I came to Princeton after the war there were no courses in relativity or gravitational physics—it was as if the field didn't exist.[6] Of course there was no course in cosmology either. So it's not too much to speak of a renaissance following a long medieval period, and we are still in the midst of this renaissance. General relativity has become physics, I mean real physics.

CCN And your own contributions to this renaissance? Which topics have you particularly worked on?

RD Oh, I'm interested in a variety of problems, some of them experimental and others of a more fundamental nature. Let me just mention two topics that are relevant to cosmology. One of them is the large dimensionless constants of nature that Dirac discussed a long time ago, suggesting that they reflect the age of the universe. Do you know about Dirac's hypothesis of varying gravity?

CCN Yes, indeed I do. In fact I interviewed him two years ago on this and related matters and he also mentioned your alternative idea.

RD Excellent, so there's no need to elaborate. Tell me, has your interview with Dirac been published?

CCN No, and it never will. It was the same kind of informal interview that we are doing now and that I mentioned I had done with Einstein a long time ago.

RD Oh yes, of course. Well, if Dirac is right, Einstein's theory of general relativity cannot be quite correct, and I also had my own reasons to think that it might only be approximately correct. I had the nagging feeling that it might not be the final answer and that Mach's principle[7] may well play a more significant role than what is usually assumed. Not only did I have this feeling, I still have it. You see, it was not at all clear whether or not general relativity incorporates some form of Mach's principle, as I wanted it to do. It's such an appealing principle.

CCN Was this the kind of consideration that led to what is called the Brans–Dicke theory of gravitation?[8]

RD Yes, it was part of it, but I had the general idea before that, including the idea that the large numbers of Dirac and Eddington should be understood from the point of view of Mach's principle. In the context of cosmology, some of the solutions of Einstein's field equations are compatible with the principle while others are not, and I thought that Mach's principle might in this way determine[9] which of the solutions correspond to physically meaningful models of the universe. Well, rather than using the Einstein equations in their original form, my student Carl Brans[10] and I suggested a more general theory that includes Einstein's as a special case. Because our equations differ slightly from those of general relativity, they also result in slightly different predictions—for example with regard to the anomalous motion of Mercury. Contrary to Einstein's theory, ours allows for an evolving gravitational constant.

CCN But in a different way from the one Dirac proposed?

RD Yes, the constant of gravitation is not a fixed constant, but one that depends on a certain parameter that allows gravity to vary in both space and time. Gravity decreases, but at a rate somewhat different from that proposed by Dirac. We soon recognized that our approach was closely related to the work done in Germany by Jordan and his collaborators,[11] although our motivations differed. I have corresponded with him, but we decided to proceed separately.

It's most interesting that our theory—and Jordan's as well—has several consequences related to geophysics and astrophysics.[12] Among the direct effects of a weakening gravitation is that the Earth would slowly expand[13]—very slowly. I have estimated an increase in its radius of about a centimetre every 500 years. Although small, over long periods of time this builds up and may have appreciable effects on the formation and distribution of land masses, but probably not enough to account for the drifting continents and the oceanic ridges. Geophysicists have recently come to a new picture of the dynamical Earth that has even been hailed as a revolutionary advance, and cosmological theory plays a surprising part in the ideas of the expanding Earth and what is known as plate tectonics.[14] Do you know what I'm referring to?

CCN Plate tectonics? I may have heard the name, but no, not really … isn't it geology?

RD It's more geophysics than geology in the classical sense. It's a kind of revival of Wegener's old idea of the drifting continents,[15] only much extended and much better justified.

CCN Ah, Alfred Wegener's theory! Yes, memories are coming back, I remember that from the old days in Germany. I actually heard a public lecture by Wegener, in Leipzig, I think, but it must have been nearly forty years ago. His theory was quite controversial and much discussed, really fashionable in some circles, but then interest faded. So Wegener has been rehabilitated, I didn't know that.

RD My point is just that we have here an interesting connection between what happens in the earth sciences and in the field of cosmology. Perhaps it's no coincidence that we are entering a new era in both sciences—but let's keep to cosmology, shall we?

CCN Absolutely. So, my question is this: did you believe in an early hot and compact state of the universe even before you became aware of the cosmic microwaves?

RD Yes I did, absolutely. I was aware of the steady-state alternative, but without taking it very seriously. On the other hand, I thought quite a lot about the origin of matter, how it might have come into existence in a natural way. Continuous creation didn't appeal to me, it seemed a pseudo-solution, something coming out of the blue. A *deus ex machina*, if you know what I mean.

CCN I do. And now, after the discovery of the cosmic microwave background, would you say that the steady-state theory is definitely ruled out?

RD Oh yes, absolutely, it's stone dead.

CCN But isn't it conceivable that the radiation can be explained in some other way,[16] perhaps within the framework of some modified version of the steady-state theory? After all, what we know is only that there is this background, we have no solid proof that it is a fossil from a past bang, isn't that true?

RD No solid proof … Hmm, I would say that we do have proof, if not as yet very solid, perhaps. I guess you're right, but in principle only. One can always keep a theory alive by means of some suitable hypothesis of an ad hoc nature. But it would be so artificial and unconvincing. Really, no one would believe in it.

CCN Do you recall when you began thinking along these lines? I mean, about a primordial universe?

RD I'm not sure, perhaps five years ago, just privately, and later I discussed these things in seminars and with my group. I didn't like the idea of a universe suddenly coming into existence,[17] out of nowhere as it were. From a philosophical point of view it makes me uncomfortable, but don't ask me why, there are people who have no problem with it. There was this Belgian who advocated a sudden birth of the universe—what was his name?

CCN You are probably thinking of Georges Lemaître and his hypothesis of a primeval atom, which he proposed in the early 1930s. He's still alive and, as far as I know, still active.

RD Right, Lemaître, of course. Now I remember. Was it really that long ago? Well, I never read his papers. In fact, I'm not as well acquainted with the cosmological literature[18] as I ought to be. My main source was Tolman's textbook[19] and later I also used Bondi's book. But I prefer to think about these problems on my own, which was the way I arrived at my ideas about the universe.

CCN Sorry, which ideas are you referring to?

RD Oh, that once there was a hot and very compact state of the universe, and that this state was not only the beginning of the expanding universe in which we live but itself the result of an earlier contracting universe—that the universe evolves in a series of huge cycles.[20] It offers an explanation of how matter and radiation came into existence without assuming some initial event that cannot be explained physically. What appeals to me is that in this way one doesn't need to explain the creation of matter. It was already there.

CCN Aha, I was a bit puzzled reading the recent paper in the *Astrophysical Journal*[21] about the oscillating universe, but now I understand better. On the other hand, you also write that the conclusion of a fossil radiation is independent of whether the universe is oscillating or not.

RD Yes, our reasoning doesn't really depend on it, it's only that it is a much more appealing assumption,[22] at least to my mind. But I must admit that my three co-authors don't share my preference for the oscillating model. Jim [Peebles] just wants a hot and compact initial state from which he can start calculating. He doesn't care where it comes from.[23] Anyway, the light in the last phase of the previous bounce would be strongly blueshifted and capable of decomposing the heavy elements into hydrogen by means of photodissociation. This was the idea, and I also calculated that the radiation in the expanding phase would be thermalized; it would have a blackbody spectrum and cool off with the expansion. A rough calculation indicated[24] an upper bound of 40 K. I didn't write it down, but I must have had it a year ago or so. Yes, about a year ago.

CCN You probably know that the idea of an oscillating or cyclic universe is an old one, and that it goes all the way back to the nineteenth century.

RD The nineteenth century? I doubt that. It can't be that old, I think, for it presupposes the notion of closed space which was only introduced with general relativity around 1920 or so. I knew the oscillating universe from Tolman's book. He discusses it at length.

CCN Yes, and Gamow had somewhat similar ideas.[25] He spoke of a "big squeeze" in which a previously existing infinite universe was compressed to almost nothing and out of which our universe emerged in the form of hot radiation and elementary particles. And just as Dr Peebles shows no interest in your idea about an oscillating universe, as you say, so Alpher and Herman ignored Gamow's idea of a single bounce. By the way, scientists and philosophers did discuss the oscillating universe decades before general relativity, but it was in a sense very different from the modern one. Just for the record.

RD I wasn't aware that Gamow was in favour of a bouncing universe. That's news to me. But in my view there was not just one such event but an indefinite number of bounces in which new matter was created at every bounce, and more and more of it. There was the classical problem of entropy increase, but that's taken care of by the new matter. Although entropy increases by a finite amount in each cycle, so does matter, and the entropy per particle of mass stays constant,[26] more or less. I know it's a speculation. As I said, it's not really necessary.

CCN But to get rid of an initial creation one must imagine an eternally cyclic universe with an infinite number of past cycles. It has been argued that this is not thermodynamically possible, first by a Dutchman by the name Zanstra[27] about a decade ago. I actually met him in Brussels in 1958, where he told me about his work. He showed that in each cycle the entropy increases by a finite amount, and for this reason the universe cannot have lived through an infinite past. There needs to have been a first cycle,[28] and then we are back at square one, aren't we?

RD Perhaps this Zanstra is right, I haven't read him. I really would prefer an eternal past. ... But then, shouldn't that be possible, I would imagine square one to be empty space and that it all started with a vacuum quantum fluctuation. I have long had an interest in the vacuum, picturing it for myself as an ether full of gravitational and electromagnetic energy,[29] and from this point of view ...

CCN I hope you don't mind me interrupting, but that doesn't solve the problem, does it? An ether or a vacuum full of energy fluctuations may be a very primitive thing, but it's still a thing and it would have to be created, wouldn't it? I mean, why this ether rather than some other?

RD Ah, now you're becoming philosophical! Perhaps one cannot explain the ultimate origin of the universe, but empty space it good enough for me, it's as close to nothingness that a physicist can come. In any case, from the point of view of physics it's not really relevant. What matters is the last bounce or just some hot and compact initial state, not necessarily initial in the strict sense. It's what some people prefer to call the "big bang" and what I like to call the primordial "fireball." It is this early phase of the universe that explains the cold microwave radiation that has just been discovered, although originally it was very hot, of course.

CCN Yes, I noticed that the article in the *New York Times*[30] uses the term "big bang." Were you already familiar with the name?

RD Jim knew it, I think, and we have used it once or twice.[31] I assume it's one of Gamow's names. "Fireball" is more to my taste,[32] it was Wheeler who came up with that suggestion.

CCN The name "big bang" was actually coined by Hoyle.

RD Really? That's strange, isn't it? I mean Hoyle doesn't believe that ...

CCN No, he doesn't, not at all. But I would like to know a little more about what happened late last year and in the early part of this year, and also of how you received news of the measurements made by Penzias and Wilson. The basic story is told by Sullivan in his fine article in the *New York Times*, but perhaps there's something to add. For example, Peebles was already working on the problem of the relic background radiation that you had previously considered, is that right?

RD Yes, I suggested that he take a closer look at the problem, which may have been last fall. As expected, he did a fine job, arriving at a present background temperature of about 10 K,[33] a much better bound than mine. He gave a couple of colloquia on his work, I believe, but didn't write it up. We were collaborating at the time on a review article that we submitted in early March, and in that article Jim reported some of his results. He also wrote a more detailed paper[34] for the *Physical Review*, but I understand that for some reason it was rejected. So our present paper is really the first one to give the proper explanation of the background radiation[35] and its connection to the primordial synthesis of helium.

CCN So, with the exception of the last paper, all this happened before you knew of the result of the Bell physicists?

RD Yes, we had a meeting with them in March,[36] actually two meetings, although only the three of us. Jim couldn't participate. Penzias and Wilson had seen a preprint of Jim's paper, or perhaps knew about it through the grapevine, and they were surprised to learn that the answer to the noise problem in their antenna was to be found in the early universe. They were not thinking cosmologically, not at all. Jim then gave a talk[37] at one of the meetings of the American Physical Society, which was the first time that the cosmological interpretation of the measured background was presented in public. It all went very fast and, of course, at the same time we were busy testing our own radiometer.

CCN Right, and then the two groups decided to publish the articles back-to-back in the *Astrophysical Journal*.

RD Well, Penzias actually proposed that we wrote a joint Bell–Princeton paper,[38] but I found that unacceptable and so we ended up with the two separate papers. But of course they were coordinated.

CCN Thank you, it's good to know the sequence of these recent events. Tell me, how certain are you that the radiation is really distributed

like that coming from a black body? After all, so far it has been measured only at a single wavelength,[39] of . . . now, what is it?

RD It's 7.4 cm, and our instrument is designed to work at 3.2 cm, but it's
not ready yet. Well, of course we need more data, but I'm confident
that future measurements will confirm the blackbody spectrum.
There's a rich field waiting for experimental astronomers. Perhaps
one day we will be able to measure the radiation in space, from rockets or even a satellite. Then we will get much better data. Who knows,
ten or twenty years from now?[40]

CCN You may well be right, with the pace of development of space
technology nowadays. Now, with regard to the Penzias–Wilson
discovery, could one say that they were just fortunate or is it an
example of the proverb that chance favours the prepared mind?[41]

RD Ah, I like that, was it Einstein who said that chance only favours
the prepared mind?

CCN As far as I know it goes back to Pasteur in the nineteenth century,
but what do you think about the discovery of Penzias and Wilson?

RD It's not up to me to judge, they certainly made a very important
discovery, no doubt about that. And yet there's something funny
about it,[42] for their minds were not really prepared for what they
found. I mean, what did they discover, really? A background noise
signal they couldn't explain—that's all. I think it's fair to say
that the recognition that the signal is a fossil radiation from the
early universe belongs to Princeton. While they discovered something, we discovered the nature of this something, which is no less
important.

    Penzias and Wilson will certainly be awarded for their discovery,
not only scientifically but also commercially. After all they are Bell
employees and the company is already using the discovery in their
advertisements. I can't help finding it slightly inappropriate. But I'm
not interested in questions of reward or priority. I just want to set the
record straight.

CCN Of course. As to priority, I assume you are aware of the much
earlier prediction of Gamow and his associates Alpher and Herman?

RD Oh yes, we actually referred to those early papers[43] in which they
discussed the synthesis of deuterium and helium in a hot primordial
universe. We're aware of them.

CCN But wait, that is not what I'm referring to. They also had the idea that as a result of the original decoupling of matter and radiation a blackbody background would be formed. Alpher and Herman were first,[44] in 1948 I think; they estimated the present temperature of the background to be about 5 K. They did so in several papers more than a decade ago.

RD Are you sure of that? A cosmic microwave background of a temperature comparable to that found by Penzias and Wilson? I can't imagine …

CCN Yes, I'm quite sure, it can be looked up. It's there, in *Nature* and also in *Physical Review*. I once had an opportunity to speak with Gamow about it. You see, in the papers that you and your group referred to there's nothing about the background radiation, so perhaps you just missed it, looking at the wrong papers.

RD Hmm, this is a little embarrassing, but … as I said, I'm not strong on the literature and Jim, I'm afraid, is no better. But now we need to dig up this old work and see what it's all about. In fact, now you mentioned it, I once attended a talk Gamow gave[45] here in Princeton. It was several years ago, perhaps in the mid-fifties, and it was about the formation of elements in the early universe. But I don't recall that he said anything about a hot initial state or of a radiation coming from it. I don't think he did.

CCN You may have forgotten about it or considered it irrelevant at the time. That's what happens. In my view, whatever it's worth, Gamow and his group deserve some share of the credit.[46]

RD Apparently they do, apparently they do. I'll discuss it with Jim.

CCN In any case, apart from the early work of Gamow and his group, no one recognized the cosmological importance of the microwave background[47] before the problem was taken up in Princeton, is that right?

RD Yes, to the best of my knowledge, but as I said …

CCN Oh, you're undoubtedly right. I have looked at the literature and found nothing. Anyway, Professor Dicke, it has been a great pleasure talking with you about this new and exciting development. Thank you so much!

RD The pleasure is mine, Mr Nelson. So this is the end of your remarkable interview project? As I told you, I find it fascinating that you have actually talked with Einstein, Eddington, and Hubble. They were among my heroes when I grew up as a teenager in St Louis.

CCN Yes, it's the end of a long journey that has brought me through half a century. It's with some sadness I admit it, for the discovery of the microwave background seems to herald a whole new chapter in the history of cosmology, and one that I would have liked to follow. Unfortunately I'm too old, but at least it has been rewarding to follow the earlier developments over all these years and a privilege to meet so many of the giants of the new science of the universe. It's almost as if I myself have been part of history, if only a very small part, of course.

Notes

Robert Henry Dicke was born on 6 May 1916 in St Louis, Missouri, and died on 4 March 1997 in Princeton, New Jersey. Following undergraduate studies at Princeton University he received his PhD in physics from the University of Rochester in 1941. Immediately thereafter he was enrolled at the Radiation Laboratory at MIT to do war-related work. He worked in particular on microwave radar, an area in which he became recognized as an authority. In 1946 he returned to Princeton, where he spent his entire academic career, since 1975 as Albert Einstein Professor of Science (see Figure 14). Exceptionally among post-World War II physicists, Dicke had a distinguished career in both experimental and theoretical physics, and he straddled the barrier that at the time made quantum mechanics and general relativity two separate worlds. In the 1950s he mostly worked on quantum optics and precision measurements of atomic structure and the interaction of matter and radiation. His work in these areas proved important for the invention of the laser. He held fifty patents, some of them on laser technology.

In the mid-1950s Dicke turned to studies of gravitational physics, first with precision experiments that verified Einstein's equivalence principle (all bodies experience the same acceleration in a gravitational field). He was a key figure in the so-called renaissance of general relativity, contributing to it experimentally as well as theoretically. Convinced of the importance of Mach's principle, in 1961, together with Carl Brans, he developed an extension of general relativity known as the scalar–tensor theory which included a gravitational constant that evolved with time. In part inspired by this theory, he turned to cosmology. Fascinated by the idea of an oscillating universe, he reasoned that during the collapse phase high-energy photons would dissociate matter and produce a fresh supply of hydrogen for a new expansion. In addition, blackbody radiation would follow. In 1964 his group at Princeton began building a radiometer to measure this radiation. Before the instrument was ready, he was informed of the detection of a cosmic microwave background by Arno Penzias and Robert

Wilson and realized that they had discovered a fossil radiation from the big bang. The paper he wrote in the *Astrophysical Journal* together with his group of three young physicists is generally recognized as the beginning of modern big bang cosmology.

Biographical sources: Lightman and Brawer (1990, pp. 201–213); Happer et al. (1999); Peebles (2008); 1978 AIP interview (see AIP Interviews).

1. *the paper that will appear shortly*: Dicke et al. (1965) was dated 7 May 1965, but actual publication was held up until 1 July and the issue of the *Astrophysical Journal* was mailed to the subscribers only in the early autumn (Peebles et al. 2009, p. 152).
2. *Radiation Lab*: The Radiation Laboratory located at the MIT (Massachusetts Institute of Technology) campus in Cambridge, MA, was founded in 1940. During World War II it played a key role in the development of microwave radar systems and related fields of microwave physics. While at the Radiation Laboratory, Dicke invented a radiometer capable of detecting microwaves from a warm body. The laboratory closed down at the end of 1945, when its main research facilities were taken over by MIT.
3. *the astronomers were simply not interested*: In an unpublished scientific autobiography of 1975, Dicke recalled that "as a very junior member of the physics department, I considered it rash to start doing astronomical research, and I could not develop any interest in the astronomy department." (Happer et al. 1999).
4. *Wheeler was around*: In the 1950s the famous Princeton physicist John Archibald Wheeler turned from nuclear physics to theoretical studies of general relativity. For the next couple of decades he was probably the leading expert in the field and its application to new areas of research, including what he in 1967 called "black holes."
5. *"renaissance of general relativity"*: From about 1925 to the mid-1950s, general relativity was cultivated by only a small minority of physicists. Compared with quantum theory, nuclear physics, and solid state physics it was a decidedly unfashionable branch of science. Its only connection to observation and experiment was cosmology, itself considered a peripheral science. The situation began to change in the late 1950s, and ten years later relativity and gravitation had become a popular area of physics with rapid theoretical as well as experimental developments. An important reason for the renaissance was new discoveries in astrophysics (such as pulsars and quasars); even more important was that new methods in experimental physics, such as the Mössbauer effect, turned Einstein's theory into a laboratory science. The development is analysed in Kaiser (1998) and Will (1993).
6. *it was as if the field didn't exist*: "Relativity and cosmology were not regarded as decent parts of physics at all," Dicke recalled in 1988. While a graduate

student at the University of Rochester he was told that general relativity "really had nothing to do with physics … [it] was a kind of mathematical discipline." Interview in Lightman and Brawer (1990, p. 204). At Princeton, there was a course in general relativity, but it was in the mathematics department. Only in 1954 did it become part of a physics course taught by John Wheeler. See Kaiser (1998).

7. *Mach's principle*: This principle, named after the Austrian physicist–philosopher Ernst Mach, is notoriously ambiguous. Einstein understood it in the sense that the space–time metric is determined by the mass of the universe, and thus that the local dynamics is determined by the universe at large. He originally believed that the relativistic cosmology of 1917 embodied Mach's principle, but later concluded that the principle could not be harmonized with the general theory of relativity. On Einstein and Mach's principle, see Pais (1982, pp. 283–287).

8. *the Brans–Dicke theory of gravitation*: For an accessible account of the Brans–Dicke (or Brans–Dicke–Jordan) theory, see Will (1993, pp. 147–159), according to whom it was "the first serious challenge to the supremacy of general relativity." The equations of this kind of gravitation theory contain a dimensionless constant, which can be chosen to fit observations. The theory is also less stringent than general relativity in the sense that it admits more solutions.

9. *Mach's principle might in this way determine*: Dicke (1959, p. 36) wrote of Mach's principle that it should "appear as a boundary condition upon solutions to Einstein's field equations, the allowed solutions being only ones which are compatible with Mach's Principle."

10. *Carl Brans*: Carl Henry Brans obtained his PhD in 1961 under Dicke's supervision. As professor of physics at Loyola University, New Orleans, he continued to do research into the mathematical aspects of general relativity.

11. *Jordan and his collaborators*: The Jordan school in gravitational physics included Engelbert Schucking, Jürgen Ehlers, and several other German theorists. See Schucking (1999) and also Chapter 12.

12. *consequences related to geophysics and astrophysics*: For a survey of these consequences, either based on Dirac's theory or the Brans–Dicke theory, see Dicke (1959) and Dicke (1962). Dicke and his Canadian student James Peebles used the temperature and age of meteorites to evaluate the two kinds of cosmologies (Dicke and Peebles 1962).

13. *the Earth would slowly expand*: "If we assume that the gravitational constant has been steadily decreasing with time, what effect would such a decrease have had upon the earth throughout its history? … Among the direct effects is the general expansion of the earth which must accompany a decrease in gravitational interaction (an increase in radius of 0.2 centimeter per century

accompanying a rate of decrease of the gravitational constant of 3 parts in 10^{11} per year)." (Dicke 1962, p. 657). As mentioned in Chapter 12, as early as 1948 Edward Teller had investigated the effects of Dirac's hypothesis upon the past climate of the Earth.

14. *what is known as plate tectonics*: The modern theory of plate tectonics emerged during the years 1960–5, primarily on the basis of advances in marine geology and palaeomagnetism. In the same period some earth scientists, including Samuel Warren Carey in Australia and Lázló Egyed in Hungary, suggested ideas of an expanding Earth in part inspired by the cosmological theories of Dirac, Jordan, and Dicke. For a summary account see Oldroyd (1996, pp. 273–278). One of the leaders of the plate tectonic revolution, the Canadian John Tuzo Wilson, adopted expansionism and considered the hypothesis of a varying gravitation "an inviting idea." See Wilson (1960, p. 882).

15. *Wegener's old idea of the drifting continents*: The German astronomer and geophysicist Alfred Wegener suggested his famous hypothesis of the drifting continents in 1912. In 1915 he presented a more elaborate version of it in his classic study *Die Entstehung der Kontinente und Ozeane* [The Origin of Continents and Oceans]. The "mobilist" hypothesis became controversial and widely known only after 1920, when it attracted much interest among scientists and lay people alike. However, by the mid-1930s it was dismissed by a majority of geologists and palaeontologists, only to be resurrected thirty years later. For Wegener's theory and its further development into plate tectonics, see, for example, Le Grand (1988).

16. *the radiation can be explained in some other way*: Together with his collaborators Jayant Narlikar and Chandra Wickramasinghe, Hoyle developed a mechanism that explained the background radiation and its spectrum without relying on a hot primordial universe. Starting in the late 1960s, they suggested that the measured microwaves were starlight thermalized by grains flowing around in interstellar space. Although they could in this way account for the cosmic microwave background within the framework of a modified steady-state theory, their explanation was dismissed by the majority of astronomers and physicists, who found it artificial, ad hoc, and unnecessary. See Kragh (1996, pp. 356–357).

17. *a universe suddenly coming into existence*: In an interview of 19 January 1988, Dicke said: "A universe that is suddenly switched on I find highly disagreeable. I guess what bothers me is a sudden barrier, a discontinuity, whether it's in time or space—because I'm used to continuity. To have space on one side of a sheet and not exist on the other I would find most disagreeable." (Lightman and Brawer 1990, p. 212). Dicke's emotional dislike of a universe with a definite beginning resembles the attitude of Eddington as mentioned in Chapter 8.

18. *I'm not as well acquainted with the cosmological literature*: Dicke's attitude to the scientific literature was shared by many of his physics colleagues, then and later. For example, Peebles recalled: "I was never strong on the literature. Still am not strong on the literature. It's so much more fun to think things through on your own than it is to read someone else's paper." (1984 AIP interview; see AIP Interviews).

19. *My main source was Tolman's textbook*: For two decades Tolman (1934) was the standard textbook in cosmology and general relativity. The other book referred to is Bondi (1952), which also included Milne's cosmological system, Dirac's varying-G cosmology, and of course the steady-state theory of the universe.

20. *the universe evolves in a series of huge cycles*: Dicke recalled: "It occurred to me, one day, that one possible way of accounting for the matter in the universe would be to have an oscillating universe, which at every collapse of the previous universe would be raised to sufficiently high temperature to get rid of all the heavy elements that were present … You would end up with nothing but protons, neutrons, mesons, electrons and what not. The universe would bounce; and in this new fresh, nascent state expand, stars would form, heavy elements would develop; and again it would bounce to some maximum size and collapse." (1985 AIP interview; see AIP Interviews).

21. *the recent paper in the Astrophysical Journal*: The seminal paper in the July 1965 issue of *Astrophysical Journal* by Dicke and his three Princeton collaborators started with the assumption of a closed oscillating universe: "The matter we see about us now may represent the same baryon content of the previous expansion of a closed universe, oscillating for all time. This relieves us of the necessity of understanding the origin of matter at any finite time in the past. In this picture it is essential to suppose that at the time of maximum collapse the temperature of the universe would exceed 10^{10} °K in order that the ashes of the previous cycle would have been reprocessed back to the hydrogen required for the stars in the next cycle." (Dicke et al. 1965, p. 415).

22. *a much more appealing assumption*: In a review paper submitted in early March 1965, at a time when they were not yet aware of the measurements of Penzias and Wilson, Dicke and Peebles discussed the "interesting embarrassment" of the origin of the universe in a singular state of infinite density. "Perhaps the most appealing possibility would be that the universe is an oscillating one," they suggested (Dicke and Peebles 1965, p. 447). When the paper appeared, it included a note added in proof referring to the results of Penzias and Wilson.

23. *He doesn't care where it comes from*: Jim Peebles confirmed that Dicke was strongly attracted to the idea of an oscillating universe: "He said so many times, 'Consider the universe as it is today, you can trace back to the universe as it

was yesterday, and then the day before. But in the conventional big bang, you run into the day zero where things stop.' So he preferred an oscillating universe." (1984 AIP interview; see AIP Interviews).

24. *A rough calculation indicated*: In the 1985 AIP interview (see AIP Interviews), Dicke recalled: "I made a rough estimate of what the temperature of the universe might be now, as the background radiation cooled off and got some 35 or 40 degrees, ... it's sort of a crude upper bound."

25. *Gamow had somewhat similar ideas*: On these ideas, see Chapter 10. In Gamow's vision, the universe had contracted for an infinite time in the past and after the bounce it would continue to expand for an infinite time in the future. Thus, it was a bouncing but not an oscillating universe.

26. *the entropy per particle of mass stays constant*: Dicke explained: "The problem of increasing entropy ... is avoided in this scheme, if you, in the process of making new matter with every collapse, you make enough new matter to permit the entropy per nucleon to stay roughly constant. Well, the implication of this is, then, every bounce of the universe is more energetic that the previous one, contains more matter, if you goes backwards in time, the universe gets smaller and smaller and finally ends up as a single quantum fluctuation or something we don't understand. But at least you're not faced with the problem of generating all that matter initially in one burst." (1985 AIP interview; see AIP Interviews).

27. *first by a Dutchman by the name Zanstra about a decade ago*: In 1957 the Dutch astrophysicist Herman Zanstra, professor of astronomy at the University of Amsterdam, published a careful analysis of cyclic universe models in which he argued that there can only be a finite number of previous cycles. A brief account is given in Kragh (2009). Zanstra was a participant of the 1958 Solvay conference in Brussels, where CCN met him.

28. *There needs to have been a first cycle*: By the 1970s it was recognized that although an oscillating universe may have an infinite future, it cannot have an infinite past. When Dicke and Peebles reconsidered the problem in 1979, they found that the entropy per nucleon increases in each cycle. "If the magnitude of the increase were the same on each cycle the universe would have to be less than 100 cycles old," they concluded. Moreover, and without answering the question: "Could the first cycle of our universe have developed out of 'nothing,' as a zero-energy quantum fluctuation leading to the production of a few quanta?" (Dicke and Peebles 1979, pp. 512–513). An oscillating universe that can be traced back to a first cycle loses much of its philosophical appeal, since the question of an absolute origin then reappears.

29. *an ether full of gravitational and electromagnetic energy*: "One suspects that, with empty space having so many properties, all that had been accomplished in destroying the ether was a semantic trick. The ether had been renamed

the vacuum." (Dicke 1959, p. 29). As mentioned in Chapter 5, the idea of a quantum vacuum-ether could be found decades earlier. Today it is associated with the dark energy discovered in the late 1990s.

30. *the article in the* New York Times: The first account of the discovery of the cosmic microwave background appeared in a front-page article in the *New York Times* of 21 May 1965 entitled "Signals Imply a 'Big Bang' Universe." "It is clear," wrote science reporter Walter Sullivan, "that Dr. Dicke and others would like to see an oscillating universe come out triumphantly. The idea of a universe born 'from nothing' raises philosophical as well as scientific problems." The following Sunday (23 May) the newspaper followed up the discovery with yet another article, "New Light Thrown on the Birth of the Universe." According to a popular article appearing in *Science News Letter* of 26 June 1965, "The 'big bang' theory holds that some 12 or more billion years ago all matter in the universe was in one place and was spewed outward in every direction by a gigantic explosion."

31. *we have used it once or twice*: In Dicke and Peebles (1965, p. 451), submitted in early March, there is a reference to "the 'big bang' theory of nucleosynthesis," which they associated with a review paper by Alpher and Herman (1953) giving a detailed account of the Gamow–Alpher–Herman theory of element formation in the early hot universe. Incidentally, the paper by Alpher and Herman did not contain the term "big bang."

32. *"Fireball" is more to my taste*: Dicke and Peebles mostly used the term to characterize the state of the universe at the time it became transparent to electromagnetic radiation. For Wheeler as the originator of "primordial fireball," see Peebles et al. (2009, p. 199).

33. *a present background temperature of about 10 K*: Peebles first presented his ongoing work at a colloquium at Wesleyan University, Connecticut, on 2 December 1964. In a later colloquium of 19 February 1965 at the Johns Hopkins Applied Physics Laboratory, Baltimore, he argued that in the case of a hot early universe the present background temperature could be as high as 10 K. He also mentioned the planned Princeton experiment of Peter Roll and David Wilkinson. Then, "Word of the experiment spread from Ken [Kenneth] Turner, who was at the Baltimore colloquium, to Bernie [Bernard] Burke, who with Turner was at the Carnegie Institution of Washington (then a center of radio astronomy), and from there to Arno Penzias and Bob Wilson at the Bell Telephone Laboratories in Holmdel." (Peebles 1993, p. 147). The main conclusions of Peebles' talk were incorporated in Dicke and Peebles (1965), received by the editors of *Space Science Reviews* on 8 March.

34. *He also wrote a more detailed paper*: The unpublished paper, received on 8 March 1965, carried the title "Cosmology, Cosmic Blackbody Radiation, and the Cosmic Helium Abundance." Although Peebles cited Alpher and

Herman (1953), he did not cite, and was not aware of, the earlier papers in which they had predicted a background radiation originating in the early universe. By early May 1965 Peebles did not know the fate of his paper. Dicke et al. (1965) included a reference to "Peebles, P. J. E. 1965, *Phys. Rev.* (in press)." See also the 1984 AIP interview with Peebles (see AIP Interviews) and Overbye (1991, p. 132), who reports Peebles' recollection: "It was rejected on the very good grounds that I hadn't looked at the history of the subject, and I should have been aware that George Gamow had plowed all this ground along with Alpher and Herman years before. I was promptly told to go look up the old stuff, and see what was new." Unknown to Peebles, his paper was refereed by Herman, who called in Alpher for help. See Alpher (2012).

35. *the proper explanation of the background radiation*: Assuming a cosmic helium abundance of 25% by weight, "we can place an upper limit on the matter density at the time of helium formation ... and hence, given the density of matter in the present universe, we have a lower limit on the present radiation temperature." (Dicke et al. 1965, p. 416). For $T = 3.5$ K they obtained a present matter density not higher than 3×10^{-32} g/cm^3, at least an order of magnitude lower than the one observed. Nonetheless, they did not consider the discrepancy damaging.

36. *we had a meeting with them in March*: According to Penzias, contact with the Princeton group was established as follows: "I mentioned our problem to Bernard Burke during a casual telephone conversation on another matter. He replied that a preprint from Princeton had come across his desk shortly before, predicting a ten-degree background at 3 cm. It was written by P. J. E. Peebles and predicted, using certain assumptions, a thermal background of radiation as a residuum of the hot, highly condensed early state of the evolution of the universe." (Penzias 1972, p. 34). The first meeting between the Princeton and Bell physicists took place at Holmdel, and it was followed with a visit by Arno Penzias and Wilson to the laboratory in Princeton.

37. *Jim then gave a talk*: See the spectrum of the cosmic background radiation that Peebles (1993, p. 147) presented in March 1965 and in which the component found by Penzias and Wilson appears quite distinct from the galactic starlight background.

38. *a joint Bell–Princeton paper*: "I remember discussing publication with Bob Dicke and suggesting a joint paper," Penzias recalled. "Dicke refused immediately, leading me then to propose a pair of back-to-back papers in *The Astrophysical Journal*." (Peebles et al. 2009, p. 151).

39. *measured only at a single wavelength*: David Wilkinson recalled that the Penzias–Wilson experiment only made the Princeton group speed up its efforts to complete their instrument. "No one would believe that what they were

seeing was heat from the big bang without measuring this spectrum … [Thermal radiation] has a very special shape. So we charged ahead in order to try and verify this spectrum at a different wavelength than they had." Interview of 25 July 2002 conducted by Michael Lemonick and reproduced in Peebles et al. (2009, pp. 200–213). Roll and Wilkinson reported their measurement at 3.2 cm in early 1966, and later the same year three more measurements were made at other wavelengths. They were all consistent with a blackbody radiation of temperature $T = 3.0 \pm 0.5$ K.

40. *ten or twenty years from now?*: It took a little longer. The famous COBE (Cosmic Background Explorer) project was initiated as early as 1972, but it took until the autumn of 1989 before the satellite was launched into orbit. The idea of looking for the cosmic microwaves by means of a satellite was mentioned as early as 1963 by Doroshkevich and Novikov (1964), and at about the same time also by Ralph Alpher. See Alpher (2012, p. 324).

41. *chance favours the prepared mind*: In a speech given at the University of Lille, France, on 7 December 1854, Louis Pasteur said, "in the fields of observation, chance only favours the prepared mind." He did not refer to his own work, but to the discovery of electromagnetism by the Danish physicist H. C. Ørsted in 1820.

42. *there's something funny about it*: Penzias and Wilson received the 1978 Nobel Prize in Physics "for their discovery of cosmic microwave background radiation." In his interview of 1985, Martin Harwit described the two physicists as "reluctant heroes," requesting a comment from Dicke. "Well, I take a philosophical view of that," he said. "In a way, one would think that pure serendipity shouldn't be so strongly rewarded, but from the standpoint of the Nobel Prize committee, what else could they do." On Harwit's suggestion that the prize could have been awarded to Alpher and Herman: "That, I would have found objectionable for some reason, I don't know why."

Peebles later recalled: "The Nobel Prize rightly went to Penzias and Wilson: they made very sure of the reality of an unexpected result, and they made sure that the world knew about it. But the Nobel committee should have included Dicke." (Peebles et al. 2009, p. 191). Peebles would himself have been a strong candidate.

43. *those early papers*: Dicke et al. (1965) referred to Alpher et al. (1948) and Alpher et al. (1953). In their slightly earlier paper (Dicke and Peebles 1965), they referred to Alpher and Herman (1953). None of the papers mentioned the cosmic microwave background.

44. *Alpher and Herman were first*: Gamow's two collaborators predicted the cosmic microwave background and its approximate temperature in a note of 1948, but their prediction was effectively forgotten. See the interview with Gamow in Chapter 10.

45. *I once attended a talk Gamow gave*: "I'd heard Gamow give a colloquium here," Dicke recalled. "He described his ideas about heavy element formation in the early universe. But the way he presented this … I distinctly remember him describing this as a completely neutron-filled universe and starting out cold, so the idea of it being hot hadn't … I didn't realize he had that idea. We should have taken this up, but just didn't." (1985 AIP interview; see AIP Interviews). According to Victor Alpher (2012, p. 321), Dicke was also in the audience at a colloquium Ralph Alpher gave on the cosmic microwave background and the formation of elements.

46. *Gamow and his group deserve some share of the credit*: This was what Gamow thought, and Alpher and Herman as well. On 29 September 1965 Gamow wrote Penzias a letter in which he complained about the neglect of his work: "The theory of, what is now knows [sic] as 'primeval fireball' was first developed by me in 1946 (Phys. Rev. *70*, 572, 1946; *74*, 505, 1948; Nature *162*, 572, 1948). The prediction of the numerical value of the present (residual) temperature could be found in Alpher & Hermann's paper (Phys. Rev. *75*, 1093, 1949) who estimate it as 5 °K. … Even in my popular book 'Creation of Universe' (Viking 1952) you can find (on p. 42) the formula $T = 1.5 \times 10^{10}/t^{\frac{1}{2}}$ °K, and the upper limit of 50 °K. Thus, you see that the world did not start with almighty Dicke." Reproduced in Penzias (1972, p. 35) and in Peebles et al. (2009, p. 154).

   Two years later the Princeton physicists privately acknowledged their failure in not being acquainted with the history of their subject. At the end of 1967 Dicke wrote a letter to Gamow in which he expressed his wish "to forget past oversights and be friends" (Kragh 1996, p. 351). Alpher and Herman did not easily forget what they felt was a lack of proper recognition of their early work. Not only were they unhappy with the attitude of Dicke, Peebles, and Penzias toward their work, they also felt that they had been unfairly treated by Gamow. See, for example, Alpher and Herman (1990). In fact, the dissatisfaction outlived the two physicists. Victor Alpher, the son of Ralph Alpher, has continued the fight for recognition of the work of his father and his collaborator Robert Herman (Alpher 2012).

47. *one recognized the cosmological importance of the microwave background*: Neither Dicke and Peebles nor Penzias and Wilson were aware of a paper published in 1964 by two Russian astrophysicists Andrei Doroshkevich and Igor Novikov. The two Russians emphasized the blackbody spectrum of the radiation and explicitly linked it to Gamow's big bang theory: "Measurements in the region [wavelength 6 to 30 cm] … are extremely important for experimental checking of the Gamow theory. … According to the Gamow theory, at the present time it should be possible to observe equilibrium Planck radiation with a temperature of 1–10 °K." They even mentioned that

measurements by means of an artificial satellite "will assist in final solution of the problem of the correctness of the Gamow theory." The paper by Doroshkevich and Novikov (1964) was submitted on 11 October 1963 and appeared in print in the spring of 1964. Later the same year it was translated into English in the widely circulated *Soviet Physics—Doklady*, but without any notice being taken of it by Western scientists. See Kragh (1996, pp. 343–344) and Novikov's recollections in Peebles et al. (2009, pp. 99–106).

AIP Interviews

Online transcripts of the extensive series of interviews collected by the Niels Bohr Library and Archives, the American Institute of Physics, can be found at <http://www.aip.org/history/ohilist/transcripts.html>. The interviews quoted in the present work are:

Bok, Bart, by David DeVorkin, 17 May 1978:
<http://www.aip.org/history/ohilist/4518_2.html>

Bondi, Hermann, by David DeVorkin, 20 March 1978:
<http://www.aip.org/history/ohilist/4519.html>

Dicke, Robert, by Martin Harwit, 18 June 1985:
<http://www.aip.org/history/ohilist/4572.html>

Dirac, Paul, by Thomas Kuhn and Eugene Wigner, April 1962 and May 1963:
<http://www.aip.org/history/ohilist/4575_1.html>

Gamow, George, by Charles Weiner, 25 April 1968:
<http://www.aip.org/history/ohilist/4325.html>

Gold, Thomas, by Spencer Weart, 1 April 1978:
<http://www.aip.org/history/ohilist/4627.html>

Peebles, James, by Martin Harwit, 27 September 1984:
<http://www.aip.org/history/ohilist/4814.html>

Sandage, Allan, by Spencer Weart, 22 May 1978:
<http://www.aip.org/history/ohilist/4380_1.html>

References

Alpher, R. A. and Herman, R. C. (1953). "The origin and abundance distribution of the elements." *Annual Review of Nuclear Science* **2**: 1–40.

Alpher, R. A. and Herman, R. C. (1972). "Memories of Gamow." In: *Cosmology, Fusion and Other Matters: George Gamow Memorial Volume* (ed. F. Reines), pp. 304–313. London: Adam Hilger.

Alpher, R. A. and Herman, R. C. (1988). "Reflections on early work on 'big bang' cosmology." *Physics Today* **41** (August): 24–34.

Alpher, R. A. and Herman, R. C. (1990). "Early work on 'big-bang' cosmology and the cosmic blackbody radiation." In: *Modern Cosmology in Retrospect* (ed. B. Bertotti, R. Balbinot, S. Bergia, and A. Messina), pp. 129–158. Cambridge: Cambridge University Press.

Alpher, R. A. and Herman, R. C. (2001). *Genesis of the Big Bang*. New York: Oxford University Press.

Alpher, R. A., Bethe, H., and Gamow, G. (1948). "The origin of chemical elements." *Physical Review* **73**: 803–804.

Alpher, R. A., Follin, J. W., and Herman, R. C. (1953). "Physical conditions in the initial stages of the expanding universe." *Physical Review* **92**: 1347–1361.

Alpher, V. S. (2012). "Ralph A. Alpher, Robert C. Herman, and the cosmic microwave background radiation." *Physics in Perspective* **14**: 300–334.

Arrhenius, G., Caldwell, K., and Wold, S. (2008). *A Tribute to the Memory of Svante Arrhenius*. Stockholm: Royal Swedish Academy of Engineering Sciences.

Arrhenius, S. (1901). "Zur Kosmogonie." *Archives Néerlandaises des Sciences Exactes et Naturelles* **6**: 862–873.

Arrhenius, S. (1903). *Lehrbuch der kosmischen Physik*. Leipzig: Hirzel.

Arrhenius, S. (1908). *Worlds in the Making: the Evolution of the Universe*. New York: Harper & Brothers.

Arrhenius, S. (1909). "Die Unendlichkeit der Welt." *Scientia* **5**: 217–229.

Baade, W. and Zwicky, F. (1934). "Remarks on super-novae and cosmic rays." *Physical Review* **46**: 76–77.

Barkan, D. K. (1999). *Walther Nernst and the Transition to Modern Physical Science*. Cambridge: Cambridge University Press.

Batten, A. H. (1994). "A most rare vision: Eddington's thinking of the relation between science and religion." *Quarterly Journal of the Royal Astronomical Society* **35**: 249–270.

Belenkiy, A. (2012). "Alexander Friedmann and the origin of modern cosmology." *Physics Today* **65** (October): 38–43.

Berendzen, R., Hart, R., and Seeley, D. (1976). *Man Discovers the Galaxies*. New York: Science History Publications.

Bernstein, J. and Feinberg, G. (eds) (1986). *Cosmological Constants: Papers in Modern Cosmology*. New York: Columbia University Press.

Birkeland, K. (1896). "Sur les rayons cathodiques sous l'action des forces magnétiques intenses." *Archives des Sciences Physiques et Naturelle* **1**: 497–512.

Birkeland, K. (1913a). "The origin of worlds." *Scientific American* **76** (Supplement): 7–9, 12, 20–22.

Birkeland, K. (1913b). *The Norwegian Aurora Polaris Expedition 1902–1903*, Vol. 1. Christiania: H. Aschehough & Co.

Blaauw, A. (1975). "Sitter, Wilhelm de." In: *Dictionary of Scientific Biography* (ed. C. C. Gillispie), Vol. 12, pp. 448–450. New York: Charles Scribner's Sons.

Blaauw, A. (1994). *History of the IAU: the Birth and First Half-Century of the International Astronomical Union*. Dordrecht: Kluwer Academic.

Bondi, H. (1948). "Review of cosmology." *Monthly Notices of the Royal Astronomical Society* **108**: 104–120.

Bondi, H. (1952). *Cosmology*. Cambridge: Cambridge University Press.

Bondi, H. (1956). "The steady-state theory of cosmology and relativity." In: *Fünfzig Jahre Relativitätstheorie* (ed. A. Mercier and M. Kervaire), pp. 152–154. Basel: Birkhäuser.

Bondi, H. (1958–1959). "Science and the structure of the universe." *Memoirs and Proceedings of the Manchester Literary Society* **101**: 58–71.

Bondi, H. (1966). "Some philosophical problems in cosmology." In: *British Philosophy in the Mid-Century* (ed. C. A. Mace), pp. 393–401. London: Allen and Unwin.

Bondi, H. (1990). *Science, Churchill & Me: the Autobiography of Hermann Bondi*. London: Pergamon.

Bondi, H. (1994). "Karl Popper (1902–1994)." *Nature* **371**: 478.

Bondi, H. and Gold, T. (1948). "The steady-state theory of the expanding universe." *Monthly Notices of the Royal Astronomical Society* **108**: 252–270.

Bondi, H. and Kilmister, C. W. (1959–1960). "The impact of 'Logik der Forschung.' " *British Journal for the Philosophy of Science* **10**: 55–57.

Boyce, N. W. (1972). "A priori knowledge and cosmology." *Philosophy* **47**: 67–70.

Brekke, A. and Egeland, A. (1983). *The Northern Light: from Mythology to Space Research*. Berlin: Springer-Verlag.

Castelvecchi, D. (2014). "Einstein's lost theory uncovered." *Nature* **506**: 418–419.

Chandrasekhar, S. (1937). "The cosmological constants." *Nature* **139**: 757–758.

Charlier, C. V. L. (1925). "On the structure of the universe." *Publications of the Astronomical Society of the Pacific* **37**: 53–76, 115–135, 177–191.

Chernin, A. D. (1994). "Gamow in America: 1934–1968." *Soviet Physics—Uspekkhi* **37**: 791–801.

Christianson, G. E. (1995). *Edwin Hubble: Mariner of the Nebulae*. New York: Farrar, Straus and Giroux.

Clerke, A. M. (1903). *Problems in Astrophysics*. London: Adam & Charles Black.

Coffey, P. (2008). *Cathedrals of Science: the Personalities and Rivalries that Made Modern Chemistry*. Oxford: Oxford University Press.

Couderc, P. (1952). *The Expansion of the Universe*. London: Faber and Faber Ltd.

Crawford, E. (1996). *Arrhenius: from Ionic Theory to the Greenhouse Effect*. Canton, MA: Science History Publications.

Crelinsten, J. (2006). *Einstein's Jury: the Race to Test Relativity*. Princeton: Princeton University Press.

De Maria, M. and Russo, A. (1989). "Cosmic ray romancing: the discovery of the latitude effect and the Compton–Millikan controversy." *Historical Studies in the Physical and Biological Sciences* **19**: 211–266.

De Sitter, W. (1917). "On Einstein's theory of gravitation and its astronomical consequences. Third paper." *Monthly Notices of the Royal Astronomical Society* **78**: 3–28.

De Sitter, W. (1930). "The expanding universe. Discussion of Lemaître's solution of the equations of the inertial field." *Bulletin of the Astronomical Institutes of the Netherlands* **5**: 211–218.

De Sitter, W. (1931a). "Some further computations regarding non-static universes." *Bulletin of the Astronomical Institutes of the Netherlands* **6**: 141–145.

De Sitter, W. (1931b). "The expanding universe." *Scientia* **49**: 1–10.

De Sitter, W. (1932). *Kosmos*. Cambridge, MA: Harvard University Press.

De Sitter, W. (1933a). [Discussion remark]. *Observatory* **56**: 182–185.

De Sitter, W. (1933b). "On the expanding universe and the time-scale." *Monthly Notices of the Royal Astronomical Society* **93**: 628–634.

Dick, S. J. (1996). *The Biological Universe: the Twentieth-Century Extraterrestrial Life Debate*. Cambridge: Cambridge University Press.

Dicke, R. H. (1959). "Gravitation—an enigma." *American Scientist* **47**: 25–40.

Dicke, R. H. (1961). "Dirac's cosmology and Mach's principle." *Nature* **192**: 440–441.

Dicke, R. H. (1962). "The Earth and cosmology." *Science* **138**: 653–664.

Dicke, R. H. and Peebles, P. J. E. (1962). "The temperature of meteorites and Dirac's cosmology and Mach's principle." *Journal of Geophysical Research* **67**: 4063–4070.

Dicke, R. H. and Peebles, P. J. E. (1965). "Gravitation and space science." *Space Science Reviews* **4**: 419–460.

Dicke, R. H. and Peebles, P. J. E. (1979). "The big bang cosmology—enigmas and nostrums." In: *General Relativity: an Einstein Centennial Survey* (ed. S. Hawking and W. Israel), pp. 504–517. Cambridge: Cambridge University Press.

Dicke, R. H., Peebles, P. J. E., Roll, P. G., and Wilkinson, D. T. (1965). "Cosmic black-body radiation." *Astrophysical Journal* **142**: 414–419.

Dieke, S. H. (1975). "Schwarzschild, Karl." In: *Dictionary of Scientific Biography* (ed. C. C. Gillispie), Vol. 12, pp. 247–253. New York: Charles Scribner's Sons.

Dingle, H. (1937). "Modern Aristotelianism." *Nature* **139**: 784–786.

Dingle, H. (1945). "Sir Arthur Eddington, O.M., F.R.S." *Proceedings of the Physical Society* **57**: 244–249.

Dingle, H. (1956). "Cosmology and science." *Scientific American* **192** (September): 224–236.

Dirac, P. A. M. (1937). "The cosmological constants." *Nature* **139**: 323.

Dirac, P. A. M. (1938). "A new basis for cosmology." *Proceedings of the Royal Society A* **165**: 199–208.

Dirac, P. A. M. (1939). "The relation between mathematics and physics." *Proceedings of the Royal Society of Edinburgh* **59**: 122–129.

Dirac, P. A. M. (1977). "Recollections of an exciting era." In: *History of Twentieth Century Physics* (ed. C. Weiner), pp. 109–146. New York: Academic Press.

Doroshkevich, A. G. and Novikov, I. D. (1964). "Mean density of radiation in the metagalaxy and certain problems in relativistic cosmology." *Soviet Physics—Doklady* **9**: 111–113.

Douglas, A. V. (1956). *The Life of Arthur Stanley Eddington*. London: Thomas Nelson and Sons.

Earman, J. (2001). "Lambda: the constant that refuses to die." *Archive for History of Exact Sciences* **55**: 189–220.

Eckert, M. and Märker, K. (eds) (2000). *Arnold Sommerfeld. Wissenschaftlicher Briefwechsel*, Vol. 1. Berlin: Verlag für Geschichte der Naturwissenschaften und der Technik.

Eddington, A. S. (1917). "Karl Schwarzschild." *Monthly Notices of the Royal Astronomical Society* **77**: 314–319.

Eddington, A. S. (1923). *The Mathematical Theory of Relativity*. Cambridge: Cambridge University Press.

Eddington, A. S. (1925). "Hugo von Seeliger." *Monthly Notices of the Royal Astronomical Society* **85**: 316–318.

Eddington, A. S. (1928). *The Nature of the Physical World*. Cambridge: Cambridge University Press.

Eddington, A. S. (1931a). "The end of the world: from the standpoint of mathematical physics." *Nature* **127**: 447–453.

Eddington, A. S. (1931b). "The expansion of the universe." *Monthly Notices of the Royal Astronomical Society* **91**: 412–416.

Eddington, A. S. (1933). *The Expanding Universe*. Cambridge: Cambridge University Press.

Eddington, A. S. (1935). *New Pathways in Science*. Cambridge: Cambridge University Press.

Eddington, A. S. (1936). *Relativity theory of Protons and Electrons*. Cambridge: Cambridge University Press.

Eddington, A. S. (1937). "The cosmical constant and the recession of the nebulae." *American Journal of Mathematics* **59**: 1–8.

Eddington, A. S. (1938). "Forty years of astronomy." In: *Background to Modern Science* (ed. J. Needham and W. Pagel), pp. 117–144. Cambridge: Cambridge University Press.

Eddington, A. S. (1939a). "Cosmological applications of the theory of quanta." In: *New Theories in Physics*, pp. 173–206. Warsaw: Scientific Collection.

Eddington, A. S. (1939b). "The cosmological controversy." *Science Progress* **34**: 225–236.

Eddington, A. S. (1939c). *The Philosophy of Physical Science*. Cambridge: Cambridge University Press.

Eddington, A. S. (1944a). "The evolution of the cosmical number." *Proceedings of the Cambridge Philosophical Society* **40**: 37–56.

Eddington, A. S. (1944b). "The recession-constant of the galaxies." *Monthly Notices of the Royal Astronomical Society* **104**: 200–204.

Eddington, A. S. (1946). *Fundamental Theory*. Cambridge: Cambridge University Press.

Egeland, A. and Burke, W. J. (2010). *Kristian Birkeland: the First Space Scientist*. Dordrecht: Springer.

Einstein, A. (1918). *Über die Spezielle und die Allgemeine Relativitätstheorie (Gemeinverständlich)*. Braunschweig: Vieweg & Sohn.

Einstein, A. (1921). "A brief outline of the development of the theory of relativity." *Nature* **106**: 782–784.

Einstein, A. (1929). "Space-time." In: *Encyclopaedia Britannica*, 14th edn., Vol. 21, pp. 105–108.

Einstein, A. (1952a). "Cosmological considerations on the general theory of relativity." In: A. Einstein et al., *The Principle of Relativity*, pp. 175–188. New York: Dover Publications.

Einstein, A. (1952b). "Do gravitational fields play an essential part in the structure of elementary particles?" In: A. Einstein et al., *The Principle of Relativity*, pp. 191–198. New York: Dover Publications.

Einstein, A. (1952c). "Foundation of the general theory of relativity." In: A. Einstein et al., *The Principle of Relativity*, pp. 109–164. New York: Dover Publications.

Einstein, A. (1979). *Autobiographical Notes*. Chicago: Open Court Publishing Company.

Einstein, A. (1982). *Ideas and Opinions*. New York: Three Rivers Press.

Einstein, A. (1993). *Collected Papers of Albert Einstein*, Vol. 5 (ed. M. Klein, A. J. Kox, and R. Schulmann). Princeton: Princeton University Press.

Einstein, A. (1998). *Collected Papers of Albert Einstein*, Vol. 8, Part A (ed. R. Schulmann, A. J. Kox, and M. Janssen). Princeton: Princeton University Press.

Einstein, A. and de Sitter, W. (1932). "On the relation between the expansion and the mean density of the universe." *Proceedings of the National Academy of Sciences* **18**: 213–214.

Eisenstaedt, J. (1989). "The early interpretation of the Schwarzschild solution." In: *Einstein and the History of General Relativity* (ed. D. Howard and J. Stachel), pp. 213–233. Boston: Birkhäuser.

Engler, G. (2005). "Einstein, his theories, and his aesthetic considerations." *International Studies in the Philosophy of Science* **19**: 21–30.

Farmelo, G. (2009). *The Strangest Man: the Hidden Life of Paul Dirac, Quantum Genius.* London: Faber and Faber.

Farrell, J. (2005). *The Day without Yesterday: Lemaître, Einstein, and the Birth of Modern Cosmology.* New York: Thunder's Mouth Press.

Fölsing, A. (1997). *Albert Einstein.* New York: Penguin Books.

Forman, P. (1971). "Weimar culture, causality, and quantum theory, 1918–1927: adaption by German physicists and mathematicians to a hostile intellectual environment." *Historical Studies in the Physical Sciences* **3**: 1–115.

Forman, P. (1973). "Scientific internationalism and the Weimar physicists: the ideology and its manipulation in Germany after World War I." *Isis* **64**: 151–180.

Frenkel, V. (1994). "George Gamow: World line 1904–1933." *Soviet Physics—Uspekhi* **37**: 767–789.

Gale, G. (2005). "Dingle and de Sitter against the metaphysicians; or, two ways to keep modern cosmology physical." In: *The Universe of General Relativity* (ed. A. J. Kox and J. Eisenstaedt), pp. 157–174. Boston: Birkhäuser.

Gamow, G. (1940). *The Birth and Death of the Sun.* New York: Viking Press.

Gamow, G. (1946a). "Expanding universe and the origin of the elements." *Physical Review* **70**: 572–573.

Gamow, G. (1946b). *Atomic Energy in Cosmic and Human Life: Fifty Years of Radioactivity.* New York: Macmillan.

Gamow, G. (1947). *One Two Three . . . Infinity: Facts and Speculations of Science.* New York: Viking Press.

Gamow, G. (1949). "On relativistic cosmogony." *Reviews of Modern Physics* **21**: 367–373.

Gamow, G. (1951). "The origin and evolution of the universe." *American Scientist* **39**: 393–407.

Gamow, G. (1952a). *The Creation of the Universe.* New York: Viking Press.

Gamow, G. (1952b). "The role of turbulence in the evolution of the universe." *Physical Review* **86**: 251.

Gamow, G. (1954a). "Modern cosmology." *Scientific American* **190** (March): 55–63.

Gamow, G. (1954b). "On the steady-state theory of the universe." *Astronomical Journal* **59**: 200.

Gamow, G. (1956a). "The evolutionary universe." *Scientific American* **192** (September): 136–154.

Gamow, G. (1956b). "The physics of the expanding universe." *Vistas in Astronomy* **2**: 1726–1732.

Gamow, G. (1961). "Gravity." *Scientific American* **204** (March): 94–106.

Gamow, G. (1962). *Gravity.* New York: Doubleday and Company.

Gamow, G. (1967). "History of the universe." *Science* **158**: 766–769.

Gamow, G. (1968). "Naming the units." *Nature* **219**: 765.

Gamow, G. (1970). *My World Line: an Informal Autobiography*. New York: Viking Press.

Gamow, G. and Fleming, J. A. (1942). "Report of the eighth annual Washington Conference on Theoretical Physics." *Science* **95**: 579–581.

Gibson, C. R. (1911). *The Autobiography of an Electron*. London: Seeley & Co.

Gingerich, O. (1994). "The summer of 1953: a watershed for astrophysics." *Physics Today* **47** (December): 34–41.

Godart, O. and Heller, M. (1978). "Un travail inconnu de Georges Lemaître." *Revue d'Histoire des Sciences* **31**: 345–359.

Goldhaber, M. (1956). "Speculations on cosmogony." *Science* **124**: 218–219.

Gregory, J. (2005). *Fred Hoyle's Universe*. Oxford: Oxford University Press.

Grossmann, E. (1925). "Hugo von Seeliger." *Astronomische Nachrichten* **223**: 297–304.

Haas, A. E. (1936). "An attempt to a purely theoretical derivation of the mass of the universe." *Physical Review* **49**: 411–412.

Haas, A. E. (1938). "Modern physics and religion." *The New Scholasticism* **12**: 1–8.

Happer, W., Peebles, P. J. E., and Wilkinson, D. T. (1999). "Robert Henry Dicke." *Biographical Memoirs, National Academy of Sciences* **77**: 1–18.

Harper, E. (2001). "George Gamow: Scientific amateur and polymath." *Physics in Perspective* **3**: 335–372.

Harper, E., Parke, W. C., and Anderson, G. D. (eds) (1997). *The George Gamow Symposium*. San Francisco: Astronomical Society of the Pacific.

Heckmann, O. (1976). *Sterne, Kosmos, Weltmodelle: Erlebte Astronomie*. Munich: Piper.

Hentschel, K. (1993). "The conversion of St. John: a case study on the interplay of theory and experiment." *Science in Context* **6**: 137–194.

Hentschel, K. (ed.) (1996). *Physics and National Socialism: an Anthology of Primary Sources*. Basel: Birkhäuser Verlag.

Hentschel, K. (1997). *The Einstein Tower: an Intertexture of Dynamic Construction, Relativity Theory, and Astronomy*. Stanford, CA: Stanford University Press.

Hertzsprung, E. (1917). "Karl Schwarzschild." *Astrophysical Journal* **45**: 285–292.

Hetherington, N. S. (1982). "Philosophical values and observations in Edwin Hubble's choice of a model of the universe." *Historical Studies in the Physical Sciences* **13**: 41–67.

Holder, R. and Mitton, S. (eds) (2013). *Georges Lemaître: Life, Science and Legacy*. Heidelberg: Springer.

Hoyle, F. (1948). "A new model for the expanding universe." *Monthly Notices of the Royal Astronomical Society* **108**: 372–382.

Hoyle, F. (1949a). "Continuous creation." *The Listener* **41**: 567–568.

Hoyle, F. (1949b). "Stellar evolution and the expanding universe." *Nature* **163**: 196–198.

Hoyle, F. (1950). *The Nature of the Universe*. New York: Harper & Brothers.

Hoyle, F. (1955). *Frontiers of Astronomy*. London: William Heinemann.

Hoyle, F. (1958). "The steady state theory." In: *La Structure et l'Évolution de l'Univers* (ed. R. Stoops), pp. 53–80. Brussels: Coudenberg.

Hoyle, F. (1959). "The relation of radio astronomy to cosmology." In: *Paris Symposium on Radio Astronomy* (ed. R. N. Bracewell), pp. 529–532. Stanford, CA: Stanford University Press.

Hoyle, F. (1961). "Observational tests in cosmology." *Proceedings of the Physical Society* **77**: 1–16.

Hoyle, F. (1981). "The big bang in astronomy." *New Scientist* **92**: 521–527.

Hoyle, F. (1982). "Steady state cosmology revisited." In: *Cosmology and Astrophysics: Essays in Honor of Thomas Gold* (ed. Y. Terzian and E. M. Bilson), pp. 17–57. Ithaca, NY: Cornell University Press.

Hoyle, F. (1990). "An assessment of the evidence against the steady-state theory." In: *Modern Cosmology in Retrospect* (ed. B. Bertotti, R. Balbinot, S. Bergia, and A. Messina), pp. 221–231. Cambridge: Cambridge University Press.

Hoyle, F. (1994). *Home is Where the Wind Blows: Chapters from a Cosmologist's Life*. Mill Valley, CA: University Science Books.

Hoyle, F. and Sandage, A. (1956). "The second-order term in the redshift-magnitude relation." *Proceedings of the Astronomical Society of the Pacific* **68**: 301–307.

Hubble, E. (1926). "Extra-galactic nebulae." *Astrophysical Journal* **64**: 321–369.

Hubble, E. (1929a). "A relation between distance and radial velocity among extra-galactic nebulae." *Proceedings of the National Academy of Sciences* **15**: 168–173.

Hubble, E. (1929b). "A clue to the structure of the universe." *Astronomical Society of the Pacific Leaflet* **23**: 93–96.

Hubble, E. (1934). *Red-Shifts in the Spectra of Nebulae*. Oxford: Clarendon Press.

Hubble, E. (1936a). *The Realm of the Nebulae*. New Haven: Yale University Press.

Hubble, E. (1936b). "Effects of red shifts on the distribution of nebulae." *Astrophysical Journal* **84**: 517–554.

Hubble, E. (1937). *The Observational Approach to Cosmology*. Oxford: Clarendon Press.

Hubble, E. (1942a). "The problem of the expanding universe." *Science* **95**: 212–215.

Hubble, E. (1942b). "The problem of the expanding universe." *American Scientist* **30**: 99–115.

Hubble, E. (1947). "The 200-inch telescope and some problems it may solve." *Publications of the Astronomical Society of the Pacific* **59**: 153–167.

Hubble, E. (1951). "Explorations in space: the cosmological program for the Palomar telescopes." *Proceedings of the American Philosophical Society* **95**: 461–470.

Hubble, E. (1953). "The laws of red-shifts." *Monthly Notices of the Royal Astronomical Society* **113**: 658–666.

Hubble, E. (1954). *The Nature of Science and Other Lectures*. San Marino, CA: Huntington Library.

Hubble, E. and Humason, M. L. (1931). "The velocity–distance relation among extra-galactic nebulae." *Astrophysical Journal* **74**: 43–80.

Hubble, E. and Tolman, R. C. (1935). "Two methods of investigating the nature of the nebular redshift." *Astrophysical Journal* **82**: 302–337.

Humason, M. L., Mayall, N. U., and Sandage, A. R. (1956). "Redshifts and magnitudes of extragalactic nebulae." *Astrophysical Journal* **61**: 97–162.

IICO (1939). *New Theories in Physics.* Warsaw: Scientific Collection.

Jago, L. (2001). *The Northern Lights: the True Story of the Man Who Unlocked the Secrets of the Aurora Borealis.* New York: Alfred A. Knopf.

Jaki, S. L. (1979). "Das Gravitations-Paradoxon des unendlichen Universums." *Sudhoffs Archiv* **63**: 105–122.

Jammer, M. (1999). *Einstein and Religion.* Princeton: Princeton University Press.

Jordan, P. (1962). "Geophysical consequences of Dirac's hypothesis." *Reviews of Modern Physics* **34**: 596–600.

Jung, T. (2005). "Franz Selety (1893–1933?): Seine kosmologische Arbeiten und der Briefwechsel mit Einstein." In: *Einsteins Kosmos* (ed. H. W. Duerbeck and W. R. Dick), pp. 125–142. Frankfurt am Main: Harri Deutsch.

Kaiser, D. (1998). "A ψ is just a ψ? Pedagogy, practice, and the reconstitution of general relativity, 1942–1975." *Studies in History and Philosophy of Modern Physics* **29**: 321–338.

Kamminga, H. (1982). "Life from space—a history of panspermia." *Vistas in Astronomy* **26**: 67–86.

Kemp, P. (ed.) (1911). *Newcomb-Engelmanns Populäre Astronomie.* Leipzig: W. Engelmann.

Kerzberg, P. (1992). *The Invented Universe: the Einstein-De Sitter Controversy (1916–17) and the Rise of Relativistic Cosmology.* Oxford: Clarendon Press.

Kienle, H. (1925). "Hugo von Seeliger." *Vierteljahrsschrift der Astronomischen Gesellschaft* **60**: 1–23.

Klee, R. (2002). "The revenge of Pythagoras: how a mathematical sharp practice undermines the contemporary design argument in astrophysical cosmology." *British Journal for the Philosophy of Science* **53**: 331–354.

Kohler, M. (1933). "Beiträge zum kosmologischen Problem und zur Lichtausbreitung in Schwerefeldern." *Annalen der Physik* **16**: 129–161.

Kragh, H. (1990). *Dirac: a Scientific Biography.* Cambridge: Cambridge University Press.

Kragh, H. (1991). "Cosmonumerology and empiricism: the Dirac–Gamow dialogue." *The Astronomy Quarterly* **8**: 109–126.

Kragh, H. (1996). *Cosmology and Controversy: the Historical Development of Two Theories of the Universe.* Princeton: Princeton University Press.

Kragh, H. (1999). "Steady-state cosmology and general relativity: reconciliation or conflict?" In: *The Expanding Worlds of General Relativity* (ed. H. Goenner et al.), pp. 377–404. Boston: Birkhäuser.

Kragh, H. (2003). "Magic number: a partial history of the fine-structure constant." *Archive for History of Exact Sciences* **57**: 395–431.

Kragh, H. (2004). *Matter and Spirit in the Universe: Scientific and Religious Preludes to Modern Cosmology.* London: Imperial College Press.

Kragh, H. (2005). "George Gamow and the 'factual approach' to relativistic cosmology." In: *The Universe of General Relativity* (ed. A. J. Kox and J. Eisenstaedt), pp. 175–188. Birkhäuser: Boston.

Kragh, H. (2007a). "Cosmic radioactivity and the age of the universe, 1900–1930." *Journal for the History of Astronomy* 38: 393–412.

Kragh, H. (2007b). *Conceptions of Cosmos. From Myths to the Accelerating Universe: a History of Cosmology*. Oxford: Oxford University Press.

Kragh, H. (2008a). *Entropic Creation: Religious Contexts of Thermodynamics and Cosmology*. Aldershot: Ashgate.

Kragh, H. (2008b). "Sitter, Willem de." In: *New Dictionary of Scientific Biography* (ed. N. Koertge), Vol. 6, pp. 455–458. Detroit: Thomson-Gale.

Kragh, H. (2008c). "Bondi, Hermann." In: *New Dictionary of Scientific Biography* (ed. N. Koertge), Vol. 1, pp. 339–243. Detroit: Thomson-Gale.

Kragh, H. (2009). "Continual fascination: the oscillating universe in modern cosmology." *Science in Context* 22: 587–612.

Kragh, H. (2011). *Higher Speculations: Grand Theories and Failed Revolutions in Physics and Cosmology*. Oxford: Oxford University Press.

Kragh, H. (2012a). "Zöllner's universe." *Physics in Perspective* 14: 392–420.

Kragh, H. (2012b). "Preludes to dark energy: zero-point energy and vacuum speculations." *Archive for History of Exact Sciences* 66: 199–240.

Kragh, H. (2012c). "Is space flat? Nineteenth-century astronomy and non-Euclidean geometry." *Journal of Astronomical History and Heritage* 15: 149–158.

Kragh, H. (2012d). "Karl Popper and physical cosmologies." *Journal for the History of Astronomy* 43: 347–350.

Kragh, H. (2013a). "The rise and fall of cosmical physics: notes for a history, ca. 1850–1920." *Arxiv*: 1304.3890 [physics.hist-ph].

Kragh, H. (2013b). "Nordic cosmogonies: Birkeland, Arrhenius and fin-de-siècle cosmical physics." *European Physical Journal H* 38: 549–572.

Kragh, H. (2013c). "Svante Arrhenius, cosmical physicist and auroral theorist." *History of Geo- and Space Sciences* 4: 61–69.

Kragh, H. (2013d). "Big bang: the etymology of a name." *Astronomy and Geophysics* 54: 2.28–22.31.

Kragh, H. (2013e). "Science and ideology: the case of cosmology in the Soviet Union, 1947–1963." *Acta Baltica Historiae et Philosophiae Scientiarum* 1: 35–58.

Kragh, H. (2013f). "'The most philosophically important of all the sciences': Karl Popper and physical cosmology." *Perspectives on Science* 21: 325–357.

Kragh, H. (2014a). "Naming the big bang." *Historical Studies in the Natural Sciences* 44: 3–36.

Kragh, H. (2014b). "The true (?) story of Hilbert's infinite hotel." *Arxiv*: 1403.0059 [physics.hist-ph].

Kragh, H. and Lambert, D. (2007). "The context of discovery: Lemaître and the origin of the primeval-atom hypothesis." *Annals of Science* **64**: 445–470.

Kragh, H. and Rebsdorf, S. (2002). "Before cosmophysics: E. A. Milne on mathematics and physics." *Studies in History and Philosophy of Modern Physics* **33**: 35–50.

Kragh, H. and Smith, R. (2003). "Who discovered the expanding universe?" *History of Science* **41**: 141–162.

Lambert, D. (2000). *Un Atome d'Univers: la Vie et l'Oeuvre de Georges Lemaître*. Brussels: Racine.

Laporte, P. M. (1956). "Cubism and relativity, with a letter of Albert Einstein." *Art Journal* **25**: 246–248.

Le Grand, H. E. (1988). *Drifting Continents and Shifting Theories*. Cambridge: Cambridge University Press.

Lemaître, G. (1929). "La grandeur de l'espace." *Revue des Questions Scientifiques* **15**: 189–216.

Lemaître, G. (1931a). "A homogeneous universe of constant mass and increasing radius." *Monthly Notices of the Royal Astronomical Society* **91**: 483–490.

Lemaître, G. (1931b). "The beginning of the world from the point of view of quantum theory." *Nature* **127**: 706.

Lemaître, G. (1934). "Evolution of the expanding universe." *Proceedings of the National Academy of Sciences* **20**: 12–17.

Lemaître, G. (1949a). "Cosmological applications of relativity." *Reviews of Modern Physics* **21**: 357–366.

Lemaître, G. (1949b). "The cosmological constant." In: *Albert Einstein: Philosopher-Scientist* (ed. P. A. Schilpp), pp. 437–456. New York: Library of Living Philosophers.

Lemaître, G. (1958a). "Rencontres avec Einstein." *Revue des Questions Scientifiques* **19**: 129–132.

Lemaître, G. (1958b). "The primaeval atom hypothesis and the problem of the clusters of galaxies." In: *La Structure et l'Évolution de l'Univers* (ed. R. Stoops), pp. 1–25. Brussels: Coudenberg.

Lemaître, G. and Vallarta, M. S. (1933). "On Compton's latitude effect of cosmic radiation." *Physical Review* **43**: 87–91.

Lepeltier, T. (2006). "Edward Milne's influence on modern cosmology." *Annals of Science* **63**: 471–481.

Lerner, E. J. (1991). *The Big Bang Never Happened*. New York: Times Books.

Leslie, J. (ed.) (1990). *Physical Cosmology and Philosophy*. New York: Macmillan.

Lightman, A. and Brawer, R. (eds) (1990). *Origins: the Lives and Worlds of Modern Cosmologists*. Cambridge, MA: Harvard University Press.

Litten, F. (1992). *Astronomie in Bayern 1914–1945*. Stuttgart: Franz Steiner Verlag.

Livio, M. (2011). "Lost in translation: mystery of the missing text solved." *Nature* **479**: 171–173.

Lovell, A. C. B. (1958). "Radio-astronomical observations which may give information on the structure of the universe." In: *La Structure et l'Évolution de l'Univers* (ed. R. Stoops), pp. 185–207. Brussels: Coudenberg.

Luminet, J.-P. (ed.) (1997). *Alexandre Friedmann, Georges Lemaître. Essai de Cosmologie.* Paris: Éditions du Seuil.

Lyttleton, R. A. (1956). *The Modern Universe.* London: Hodder and Stoughton.

McCrea, W. H. (1950). "The steady-state theory of the expanding universe." *Endeavour* 9: 3–10.

McVittie, G. C. (1967). "Georges Lemaître." *Quarterly Journal of the Royal Astronomical Society* 8: 294–297.

Mayall, N. U. (1970). "Edwin Powell Hubble." *Biographical Memoirs, National Academy of Sciences* 41: 175–214.

Meyenn, K. (ed.) (2011). *Eine Entdeckung von Ganz Ausserordentlicher Tragweite: Schrödinger's Briefwechsel*, Vol. 2. Heidelberg: Springer.

Miller, A. I. (2001). *Einstein, Picasso: Space, Time and the Beauty that Causes Havoc.* New York: Basic Books.

Milne, E. A. (1938). "On the equations of electromagnetism." *Proceedings of the Royal Society of London A* 165: 313–357.

Mitton, S. (2005). *Fred Hoyle: a Life in Science.* London: Aurum Press.

Moszkowski, A. (1921). *Einstein: Einblicke in seine Gedankenwelt.* Hamburg: Hoffmann und Campe.

Munns, R. P. D. (2013). *A Single Sky: How an International Community Forged the Science of Radio Astronomy.* Cambridge, MA: MIT Press.

Newcomb, S. (1906). *Side-Lights on Astronomy: Essays and Addresses.* New York: Harper and Brothers.

North, J. (1990). *The Measure of the Universe: a History of Modern Cosmology.* New York: Dover Publications.

Norton, J. (1999). "The cosmological woes of Newtonian gravitation theory." In: *The Expanding Worlds of General Relativity* (ed. H. Goenner et al.), pp. 271–324. Boston: Birkhäuser.

Nussbaumer, H. (2014). "Einstein's conversion from his static to an expanding universe." *European Physical Journal H* 39: 37–62.

Nussbaumer, H. and Bieri, L. (2009). *Discovering the Expanding Universe.* Cambridge: Cambridge University Press.

Oldroyd, D. (1996). *Thinking about the Earth: a History of Ideas in Geology.* London: Athlone Press.

O'Raifeartaigh, C. and McCann, B. (2014). "Einstein's cosmic model of 1931 revisited: an analysis and translation of a forgotten model of the universe." *European Physical Journal H* 39: 63–86.

O'Raifeartaigh, C., McCann, B., Nahm, W., and Mitton, S. (2014). "A steady-state model of the universe by Albert Einstein." *Arxiv*: 1402.0132 [physics. hist-ph].

Osterbrock, D. E. (1992). "The appointment of a physicist as the director of the astronomical center of the world." *Journal for the History of Astronomy* **23**: 155–165.

Osterbrock, D. E. (2001). *Walter Baade: a Life in Astrophysics*. Princeton: Princeton University Press.

Osterbrock, D. E. and Rogerson, J. B. (1961). "The helium and heavy-element content of gaseous nebulae and the sun." *Publications of the Astronomical Society of the Pacific* **73**: 129–134.

Overbye, D. (1991). *Lonely Hearts of the Cosmos: the Scientific Quest for the Secret of the Universe*. New York: HarperCollins.

Pais, A. (1982). *Subtle is the Lord: the Science and Life of Albert Einstein*. New York: Oxford University Press.

Paul, E. R. (1986). "J. C. Kapteyn and the early twentieth-century universe." *Journal for the History of Astronomy* **17**: 155–182.

Paul, E. R. (1993). *The Milky Way Galaxy and Statistical Cosmology, 1890–1924*. Cambridge: Cambridge University Press.

Peebles, P. J. E. (1993). *Principles of Physical Cosmology*. Princeton: Princeton University Press.

Peebles, P. J. E. (2008). "Dicke, Robert Henry." In: *New Dictionary of Scientific Biography* (ed. N. Koertge), Vol. 2, pp. 280–284. Detroit: Thomson-Gale.

Peebles, P. J. E. (2014). "Discovery of the hot big bang: what happened in 1948." *European Physical Journal H* **39**: 205–223.

Peebles, P. J. E., Page, L. A., and Partridge, R. B. (2009). *Finding the Big Bang*. Cambridge: Cambridge University Press.

Penzias, A. A. (1972). "Cosmology and radio astronomy." In: *Cosmology, Fusion and Other Matters: George Gamow Memorial Volume* (ed. F. Reines), pp. 29–47. London: Adam Hilger.

Peratt, A. L. (1985). "Birkeland and the electromagnetic cosmology." *Sky and Telescope* **69**: 389–391.

Peruzzi, G. and Realdi, M. (2011). "The quest for the size of the universe in early relativistic cosmology (1917–1930)." *Archive for History of Exact Sciences* **65**: 659–689.

Plaskett, J. S. (1933). "The expansion of the universe." *Journal of the Royal Astronomical Society of Canada* **27**: 235–252.

Poincaré, H. (1892). "Non-Euclidean geometry." *Nature* **45**: 404–407.

Poincaré, H. (1913). *Dernières Pensées*. Paris: Flammarion.

Pyenson, L. (1982). "Relativity in late Wilhelmian Germany: the appeal to a preestablished harmony between mathematics and physics." *Archive for History of Exact Sciences* **27**: 137–155.

Reines, F. (ed.) (1972). *Cosmology, Fusion and other Matters: George Gamow Memorial Volume*. London: Adam Hilger.

Riesenfeld, E. H. (1931). *Svante Arrhenius*. Leipzig: Akademische Verlagsgesellschaft.

Robertson, H. P. (1933). "Relativistic cosmology." *Reviews of Modern Physics* **5**: 62–90.

Robertson, H. P. (1936). "Relativity—20 years after." *Scientific American* **160** (June): 358–359; **161** (July): 22–24.

Roseveare, N. T. (1982). *Mercury's Perihelion. From Le Verrier to Einstein.* Oxford: Clarendon Press.

Rowe, D. E. (2001). "Einstein meets Hilbert: at the crossroads of physics and mathematics." *Physics in Perspective* **3**: 379–424.

Roxburgh, I. W. (2007). "Hermann Bondi." *Biographical Memoirs of Fellows of the Royal Society* **53**: 45–61.

Rypdal, K. and Brundtland, T. (1997). "The Birkeland terrella experiments and their importance for the modern synergy of laboratory and space plasma physics." *Journal de Physique IV* 7, C4: 113–132.

Sagan, C. (1997). *The Demon-Haunted World: Science as a Candidate in the Dark.* London: Headline.

Sandage, A. R. (1995). "Practical cosmology: Inventing the past." In: *The Deep Universe* (ed. A. R. Sandage, R. G. Kron, and M. S. Longair), pp. 1–23. Berlin: Springer-Verlag.

Sandage, A. R. (1998). "Beginnings of observational cosmology in Hubble's time: historical overview." In: *The Hubble Deep Field* (ed. M. Livio, S. M. Fall, and P. Madau), pp. 1–26. Cambridge: Cambridge University Press.

Schemmel, M. (2005). "An astronomical road to general relativity: the continuity between classical and relativistic cosmology in the work of Karl Schwarzschild." *Science in Context* **18**: 451–478.

Schmeidler, F. (1975). "Seeliger, Hugo von." In: *Dictionary of Scientific Biography* (ed. C. C. Gillispie), Vol. 12, pp. 282–283. New York: Charles Scribner's Sons.

Schnippenkötter, J. (1920). *Das Entropiegesetz: Seine Physikalische Entwicklung und seine Philosophische und Apologetische Bedeutung.* Essen: Fredebeul & Koenen.

Schucking, E. L. (1999). "Jordan, Pauli, politics, Brecht, and a variable gravitational constant." *Physics Today* **52** (October): 26–31.

Schwarzschild, K. (1900). "Über da zulässige Krümmungsmass des Raumes." *Vierteljahrschrift der Astronomischen Gesellschaft* **35**: 337–347.

Schwarzschild, K. (1992). *Gesammelte Werke* (ed. H. H. Voigt), 3 Vols. Berlin: Springer-Verlag.

Seeley, D. and Berendzen, R. (1972). "The development of research in interstellar absorption, c. 1900–1930." *Journal for the History of Astronomy* **3**: 52–64, 75–86.

Seeliger, H. (1895). "Über das Newtonsche Gravitationsgestz." *Astronomische Nachrichten* **137**: 129–136.

Seeliger, H. (1898). "On Newton's law of gravitation." *Popular Astronomy* **5**: 544–551.

Seeliger, H. (1906). "Das Zodiakallicht und die empirischen Glieder in der Bewegung der inner Planeten." *Bayerischen Akademie, Sitzungsberichte, Math.-Phys. Klasse* 595–622.

Seeliger, H. (1909). "Über die Anwendung der Naturgesetze auf das Universum." *Bayerischen Akademie, Sitzungsberichte, Math.-Phys. Klasse* 3–25.

Seeliger, H. (1913). "Bemerkungen über die sogenannte absolute Bewegung, Raum und Zeit." *Vierteljahrschrift der Astronomischen Gesellschaft* 48: 195–201.

Seeliger, H. (1916). "Über die Gravitationswirkung auf die Spektrallinien." *Astronomische Nachrichten* 202: 83–86.

Seeliger, H. (1920). "Untersuchungen über das Sternsystem." *Bayerischen Akademie, Sitzungsberichte, Math.-Phys. Klasse* 87–144.

Singh, J. (1970). *Great Ideas and Theories of Modern Cosmology.* New York: Dover Publications.

Smith, R. W. (1982). *The Expanding Universe: Astronomy's 'Great Debate' 1900–1931.* Cambridge: Cambridge University Press.

Smith, R. W. (1990). "Edwin P. Hubble and the transformation of cosmology." *Physics Today* 43 (April): 52–58.

Smith, R. W. (2006). "Beyond the big galaxy: the structure of the stellar system 1900–1952." *Journal for the History of Astronomy* 37: 307–342.

Soddy, F. (1909). *The Interpretation of Radium.* London: John Murray.

Spencer Jones, H. (1935). "Willem de Sitter." *Monthly Notices of the Royal Astronomical Society* 95: 343–347.

Stanley, M. (2007). *Practical Mystic: Religion, Science, and A. S. Eddington.* Chicago: University of Chicago Press.

Stoffel, J.-F. (ed.) (1996). *Mgr. Georges Lemaître, Savant et Croyant.* Louvain: Centre Interfacultaire d'Étude en Histoire des Sciences.

Struve, O. and Zebergs, V. (1962). *Astronomy of the 20th Century.* New York: Macmillan.

Sullivan, W. T. (1990). "The entry of radio astronomy into cosmology: Radio stars and Martin Ryle's 2C survey." In: *Modern Cosmology in Retrospect* (ed. B. Bertotti, R. Balbinot, S. Bergia, and A. Messina), pp. 309–330. Cambridge: Cambridge University Press.

Teller, E. (1948). "On the change of physical constants." *Physical Review* 73: 801–802.

Tolman, R. C. (1934). *Relativity, Thermodynamics, and Cosmology.* Oxford: Oxford University Press.

Toulmin, S. (1962). "Historical inference in science: geology as a model for cosmology." *The Monist* 47: 142–158.

Trabert, W. (1911). *Lehrbuch der Kosmischen Physik.* Leipzig: B. G. Teubner.

Tropp, E. A., Frenkel, V. Y., and Chernin, A. D. (1993). *Alexander A. Friedmann: the Man Who Made the Universe Expand.* Cambridge: Cambridge University Press.

Ulam, S. M. (1972). "Gamow—and mathematics." In: *Cosmology, Fusion and other Matters: George Gamow Memorial Volume* (ed. F. Reines), pp. 272–279. London: Adam Hilger.

Walker, A. G. (1933). "Distance in an expanding universe." *Monthly Notices of the Royal Astronomical Society* **94**: 159–167.

Walter, S. (ed.) (2007). *La Correspondance entre Henri Poincaré et les Physiciens, Chimistes et Ingénieurs*. Basel: Birkhäuser.

Watson, J. D. (2002). *Genes, Girls, and Gamow: After the Double Helix*. New York: Random House.

Way, M. J. and Hunter, D. (eds) (2013). *Origins of the Expanding Universe 1912–1932*. San Francisco: Astronomical Society of the Pacific.

Weston Smith, M. (2013). *Beating the Odds: the Life and Times of E. A. Milne*. London: Imperial College Press.

Whitrow, G. J. (1959). *The Structure and Evolution of the Universe*. New York: Harper.

Whitrow, G. J. and Bondi, H. (1954). "Is physical cosmology a science?" *British Journal for the Philosophy of Science* **4**: 271–283.

Whittaker, E. T. (1946). *Space and Spirit: Theories of the Universe and the Arguments for the Existence of God*. London: Thomas Nelson and Sons.

Will, C. M. (1993). *Was Einstein Right? Putting General Relativity to the Test*. Oxford: Oxford University Press.

Wilson, J. T. (1960). "Some consequences of expansion of the Earth." *Nature* **185**: 880–882.

Worrall, J. (1982). "The pressure of light: the strange case of the vacillating 'crucial experiment'." *Studies in History and Philosophy of Science* **13**: 133–171.

Yourgrau, W. (1970). "The cosmos of George Gamow." *New Scientist* **48**: 38–39.

Zel'dovich, Y. B. and Novikov, I. D. (1983). *Relativistic Astrophysics, II: the Structure and Evolution of the Universe*. Chicago: University of Chicago Press.

Zimmerman, E. J. (1955). "Numerical coincidences in microphysics and cosmology." *American Journal of Physics* **23**: 136–141.

Index